Bill Reynolds

FERRARI
The World's Most Exotic Sportscar

Bill Reynolds

FERRARI
The World's Most Exotic Sportscar

MAGNA BOOKS

Featuring the photography of Nicky Wright

CLB 2984
This 1993 edition published by Magna Books
Magna Road, Wigston, Leicester LE18 4ZH
© 1993 Colour Library Books Ltd., Godalming, Surrey
All rights reserved
Printed and bound in Hong Kong
ISBN 1 85422 563 4

The Origin of the Legend

Very few people make ultra-fast, ultra-expensive, race-winning cars as a purely business proposition; there are too many other ways to make money, most of them easier and safer. There has to be a passion, a belief in the extraordinary, and (typically) a monstrous ego, if there is to be the drive and energy to build something truly remarkable.

Ferraris are truly remarkable. At their worst, they have been overpowered rust-buckets; but at their best … Ah, at their best ….

Enzo Ferrari, for whom the *marque* was named, was born in Modena in 1898 and was a successful racing driver by 1919. In 1920 he joined Alfa Romeo as a test driver, and came second in that year's Targa Florio. In 1929, *Scuderia Ferrari* was founded to modify and race Alfa Romeos; his *scuderia* (racing stable) bore as its symbol the prancing horse

As early as 1934-35, Ferrari had crossed the line from modifying cars (and racing them very successfully) to building cars; the twin-engined *Bimotore* cars used two straight-eight Alfa Romeo engines, one in front of the driver and the other behind, and funneled both through a common clutch and drive-train. Ferrari-built racers were prepared both for the *scuderia* and for customers.

The *scuderia* was enormously successful, but suddenly in the late 1930s it found that it was at a considerable disadvantage next to Mercedes-Benz and Auto Union, largely as a result of Hitler pumping enormous sums of money into the German racing effort as a public-relations exercise. *Il Duce* also supported racing enthusiastically, but without anything like the finance.

In 1938, Alfa Romeo took over the *scuderia* (with Ferrari as manager) and ran it as Alfa Corse, but in 1939 Ferrari chafed at being an employee and left Alfa to set up a company making machine tools. Then, also in 1939, he agreed to build a racer for Alberto Ascari and the Marchese Machiavelli for the Mille Miglia. It was a straight-eight based around two Fiat cylinder heads mounted on a specially-cast block. The car retired from the race – the last Mille Miglia before the Second World War – but showed promise. It was never developed, though, because Ferrari's factory perforce switched to war *matériel*.

When the war was over, *Commendatore* Ferrari went back to cars. He did it, though, in a unique way.

The first Ferraris were either sports-racers or out-and-out Grand Prix racers, supported by the machine-tool business but essentially expected to pay their own way. This they were to do by sales to privateers, and by winning races and the purses that went with the wins. Only later did Ferrari begin to build cars which were intended primarily for road use, and even then the emphasis was on *fast* road use.

The *Tipo* 125 of 1946-1947 took its name from the 125cc capacity of each cylinder, of which there were twelve, for a total swept volume of only 1,500cc. One-and-a-half-liter V12 engines are clearly not for mass-market saloons, and in fact only three cars were built in 1947, followed by nine in 1948 and 30 in 1949.

For 1948 the capacity of the engine rose to two liters (166cc per cylinder, hence *Tipo* 166). The car's racing debut was at the Targa Florio, which it won, and then the same car won the Mille Miglia a month later. A *Tipo* 166 was even raced at the Monaco Grand Prix, though it crashed on lap 58. Meanwhile, the 125 engine in super-charged form (with an alleged 225 bhp at 7,000 rpm) was being raced in the Grand Prix, but not actually winning – though we shall return to out-and-out racers later.

From this emphasis on competition, though, it was clear that Ferrari's ambition was to build very quick motor cars, and to sell them to a discerning and wealthy clientele; which begs the question, just how did the Maranello firm do it?

Anatomy of a Ferrari

Any Ferrari, from the original racers to the latest models, can be regarded as a mixture of three ingredients. In clear order of importance they are the engine, the chassis/running gear, and the (lightweight) body.

The engine is what any Ferrari is about, and the engines have historically been very, very strong and very, very powerful. All too often, a powerful engine is fragile and requires cosseting and frequent rebuilds, but Ferraris are not like that. If you look after a Ferrari engine – change the oil, and keep the valve clearances within tolerances – it is an astonishingly reliable piece of machinery.

The *Tipo* 166 came with a minimum power output of about 110 bhp (55 bhp/liter) and a maximum of about 150 bhp (75 bhp/liter). Although these figures are not too remarkable by the standards of the 1990s, we are not talking about the 1990s; we are talking about the 1940s, when figures of 25-30 bhp/liter were regarded as normal: Chevrolet's 3,859cc straight-six delivered 105 bhp, for example, a tad less power than the Ferrari from an engine almost twice the size.

Ferrari's V12 achieved its power in a number of ways. To begin with, a large number of cylinders translates into smaller reciprocating masses and a higher engine speed – which means, of course, that there are lots of power strokes per minute. Overhead cams, one on each bank, reduced valve-train inertia, which meant that valve float was not really an issue. The valves themselves were short-stemmed (to reduce mass) and controlled by twin hairpin springs; twin for more positive closure, hairpin for improved cooling and reduced engine height. There were only two valves per cylinder, though other manufacturers had tried four-valve heads (especially for motorcycles, which Scuderia Ferrari had also raced) long before World War II.

Both the block and the heads were light alloy; liners were cast iron. The six-throw, seven-bearing crankshaft rode in conventional anti-friction bearings, which were soon replaced with Vandervell Thinwall bearings.

The chassis and running gear were distinctly primitive next to the engine. The crash (non-synchro) gearbox was enormously strong, but it had to be; drivers were specifically enjoined (as they are to this day) not to use the clutch any more than they had to, and never to slip it for fear of cooking it. Slam-bang power take-up was the norm. The clutch on Ferraris has traditionally been heavy, and on non-servo brakes fitted with hard competition linings (which needed to be warmed up before they would work properly), brake effort was traditionally high, too.

The gearbox fed the power to a live rear axle – a feature that Ferrari was to retain long after almost everyone else had gone to independent rear suspension – and the whole plot was tied together with a light, but conventional, ladder chassis made of oval-section steel tubing. Some early Ferraris, and some of the later ones too for that matter, did not handle as well as they should have, but that has always been true of any very powerful car. Those cars which enjoy legendary handling are rarely stressed to the limit; at best they are racing chassis that can handle far more power than could be supplied by the down-rated engines with which they are fitted.

So far, the whole package was very much a driver's car, with taut handling and plenty of power. The bodies were not always such good news.

They were supplied by any number of coach builders, most of whom worked by the traditional method of bashing sheet-metal with large, flat-faced hammers, using tree-stumps as formers. This is a skill which goes back centuries – medieval and earlier armor was hammered out by a very similar process – and it has the advantage that a skilled panel-beater can produce almost any shape with a surprising degree of accuracy, often creating something that is very close to a work of art in the process.

The drawbacks of the technique, however, are that repeatability is problematical (precise dimensions are most assuredly not reproducible) and that quality, especially of steel bodies, can be very variable. Fitting body panels is often a question of "ease to fit," and convincingly rust-proofing large numbers of hand-bashed steel panels would be difficult even if the body-builder bothered, which they often didn't. In the worst cases, the hidden side of the panel would not be prepared at all, but merely left as bare metal.

To make matters worse, the way in which the panels were put together was often casual to the point of irresponsibility. Dissimilar metals were casually riveted together or connected with bolts (or self-tapping screws) of yet a third metal, which was an invitation to electrolytic corrosion. In a warm, dry climate, where the car might well be driven only in dry weather, none of this is necessarily a disaster, but even an accidental exposure to rain can cause nightmares a few years or even a few months down the road. In Italy or southern California, Ferraris last a lot better than anywhere else.

No matter how good or bad the bodies were, though, they were usually extremely light. Indeed, the whole car was usually extremely light; not until the America of 1950 did a Ferrari exceed a tonne (2,200 lb.) in weight, and most cars from the period were in the 2,000 lb. (900 kg) range. Weights could vary widely, though, according to the body style chosen, the materials used (steel or light alloy) and the number and quantity of accessories, including such basics as sound insulation.

Light weight and a powerful motor is a potent combination. The crudest measure of its value comes from the power-to-weight ratio, which Americans sometimes express in pounds per brake horsepower, but which the rest of the world typically expresses in brake horsepower per ton or tonne. In this book I have used the metric tonne, 2,200 lb. or 1,000 kg, as a handy and

entirely appropriate compromise between the standard or long ton (2,240 lb.) and the American or short ton (2,000 lb.). I have also used manufacturers' claimed horsepower, which vary widely in honesty (in general, the newer the car, the greater the honesty) and which may have been derived in a number of ways: principally "gross" (with no engine ancillaries and little or no silencing) and "nett" (with some or all engine ancillaries fitted and a muffler). Bear these discrepancies in mind whenever you read about power-to-weight ratios, or indeed about raw power. Old SAE figures were gross, and typically 10-15 per cent more than DIN, while new SAE figures are nett and are typically 5-10 per cent less than DIN, all for the same power output!

A Ferrari 166 Inter offered 122 bhp/tonne to 167 bhp/tonne; for comparison, the original (and roughly contemporary) Porsche 356 offered 52 bhp/tonne, while the first Jaguar XK120 offered 120 bhp/tonne in 1949. A modern Ferrari F40 offers rather over 420 bhp/tonne; a modern Yugo compact has up to about 80 bhp/tonne.

Of course, you can always compensate for more power by adding more weight, which is why power-to-weight ratios do not tell the whole story. A two-tonne (4,400 lb.) car with a 1,000 bhp Spitfire engine has 500 bhp/tonne; but it also has immense inertia. This makes it hard to change direction at a corner; hard to stop, with tremendous amounts of heat to be dissipated from the brakes; unresponsive to the road surface, because the suspension has to be stiff to support that much weight; and very, very thirsty because of the sheer amount of metal you are moving about.

So, the pattern of the Ferrari was set early on as a superb engine in a good (but not necessarily outstanding) chassis clothed in a body which varied widely in build quality, but which was light. The logical thing to do, therefore, is to look at the development of the mainstream of Ferrari design for many years, the V12 motor, and to see what sort of automobiles it was used to power.

The Heart of the Matter

The original Type 166 was soon bored an extra 0.2 inch or 5 mm to increase the swept volume to 2,341cc, or 195cc per cylinder, followed by yet another overbore to 2.68 inch (68 mm) for 212cc per cylinder and a total swept volume of 2,562cc; the new cars were accordingly the Type 195 and the Type 212.

As you might expect, customers could specify the degree of power they wanted; the less powerful engines were more tractable and longer lived, while the more powerful engines were increasingly "peaky" and demanded more and more care and attention; clearly, a single twin-choke Weber is easier to set up than three twin-choke 36DCF Webers. The Type 195 delivered anything from 130 bhp to 180 bhp, while the Type 212 offered 130 bhp to 170 bhp with rather less effort. The weight of the cars went up to about 2,100 lb. (950 kg), but the power-to-weight ratios were still generous: anything from 137 bhp/tonne to 190 bhp/tonne. Production of cars with these engines was still tiny – maybe 80 of the basic 212 Inter model (at $9,500 and up) and a couple of dozen of the more powerful 212 export, which could hit 140 mph (225 kph). Bodies came from Vignale, Allemano, Bertone, Ghia, Pinin Farina, Touring, and more, but the definitive body style was the *barchetta* or "little boat" open roadster by Touring. Touring also made a *berlinetta* coupé.

Because the engine was very nearly as much a custom item as the body, even the swept volume could be varied to suit the buyer's requirements, and 1951 saw a redesign of the V12 motor with a bigger block, the so-called "Lampredi" engine as distinct from the "Colombo" (both being named after their designers). The first Lampredi motor was the Type 340, which predictably displaced 4.1 liters (12x340=4,080cc) and which was also made available as the Type 342 with the same displacement (there was always a good deal of rounding!), slightly less horsepower but rather more tractability. The 342 developed 200 bhp, but the most powerful 340 Mexico motor was good for about 280 bhp. Then in 1953 the Lampredi motor was enlarged to 4,523cc for the Type 375, which delivered 300 bhp in touring form (when it might be asked to move as much as 2,750 lb., 1,250 kg) of motor car, or up to 340 bhp for competition – in which case weight might fall as low as a tonne.

The Lampredi engine was extremely unusual in that it was designed from the start to be run on "dope" – special racing fuels with high thermal content but requiring a high compression ratio as well. Very high compression ratios, well in excess of 10:1, were available by choosing different Borgo pistons, and this would have blown gaskets on the Colombo engine, where the cylinder head bolts were barely up to the job. On the Lampredi engine, therefore, the wet liners were screwed into the cylinder head and sealed at their bottom end with rubber rings.

7

The cars with the new, big-block engine all received the appellation "America" – Type 340 America, Type 342 America, etc. – presumably because Americans traditionally have a liking for bigger engines. As before, bodies were made by all kinds of people and production numbers were tiny: 22 of the Type 340, half a dozen Type 342s, and a dozen Type 375s, spread between 1951 to 1955. The earlier, small-block models also continued in production, so anyone who could afford a Ferrari was spoiled for choice. Depending on the engine and the state of tune, maximum speeds ranged from about 110-120 mph (about 180-200 kph) for the old 166 models through 120-150 mph (200-240 kph) for the various non-racing America models up to a claimed 174 mph (280 kph) for the 340 Mexico, which was built with the Carrera Panamericana in mind. A racing 375 America, the 375 Plus Roadster, won both the 1954 Le Mans and the 1954 Carrera Panamericana.

When you consider that Le Mans is a *Grand Prix de l'Endurance* and that the Carrera Panamericana ran the full length of Mexico, you can see that these were real sports-racers, which could hold their own in the most demanding road races in the world while still remaining very close to the road-going models. The bodies varied enormously, from purposeful and indeed brutal roadsters with twin aero screens through to elegant and well-balanced coupés which might even have 2+2 bodies – though the "+2" in the back would rapidly find themselves uncomfortable on anything but the shortest journey.

These America cars were also less than easy to drive, especially if you wanted "full chat" racing power. The five-speed gear change is most politely described as recalcitrant; the brakes and clutch require Sumo-style thigh muscles; and until you got up to a reasonable speed, the steering was extremely heavy. The interior noise was considerable; heat from the engine cooked you unless you opened a window; and the ventilated drum brakes faded quickly. In a word, they were pretty brutal.

Meanwhile, though, the 250 Europa appeared with a small version of the Lampredi engine – only about 21 of these 3-liter cars were built – while the Colombo engine made a big comeback in 1954 in the 250GT.

In many ways, the 250GT marked Ferrari's coming of age. The 250GT was bored to no less than 2.87 inch (73 mm) while remaining at the old 2.31 inch (58.8 mm) stroke, and delivered anything from 200 to 220 bhp to propel a car which might weigh anything from 2,600 lb. to 2,900 lb. – say 1,250 kg to 1,400 kg.

The increase in weight tells the story. At last, Ferrari had decided to make a road-going car with some concessions to passenger comfort, and the Americans in particular loved it. To purists, the new Ferraris might have seemed overweight and underpowered, though this gives some idea of what people had come to expect from Ferrari: 143 bhp/tonne to 176 bhp/tonne is still significant, and indeed compares with the contemporary Corvette, which had just acquired a 195 bhp V8 to propel its 2,700 lb. (1,227 kg – 159 bhp/tonne). Top speed depended on gearing: for the American market, where acceleration was the thing, low-geared rear axles kept the top speed down to 125 to 130 mph (200 to 210 kph), while geared for *autostrade* the car was good for over 150 mph (240 kph). The 48/52 front/rear weight balance was superb. Most of the 2,500 or so 250GT models (production ceased in the early 1960s) were bodied by Pinin Farina and Vignale, usually as coupés, with prices of around $10,000 in the United States. The rear seats in most "2+2" designs were vestigial, and indeed in many cases they were only luggage shelves.

They were strange cars. The frequent firing impulses of the V12 engine meant that tickover could be very slow, as little as 600 rpm, and that very little flywheel mass was required: the rate at which a Ferrari V12 can pick up speed (and lose speed between gear changes!) has to be experienced to be believed. The V12 also meant that the engine was enormously flexible; a *Sports Car Illustrated* road test reckoned that you could go from 10 mph to 127 mph (their one-way maximum) in top gear. The gearbox was full-synchro, the clutch remarkably light, and the whole thing was extraordinarily civilized – until you started driving fast, when the engine started screaming as only a relatively small V12 can scream, and the car doggedly refused to demonstrate even reasonable vices. Of course it was noisy, and no one could call the ride "boulevard," but all the contemporary road-testers agreed that it was wonderful. A lot depended on what one was comparing it with, but it was much easier to drive than the bigger-engined cars even if it did have the same bizarre, reversed gearshift gate with first and second on the right and second and third on the left.

Today, it is still a fascinating car, especially if one gets one of the later models when power had risen to around 300 bhp, but it shows its age. The wheels are narrow,

so it is comparatively easy to steer with the throttle, and the interior is curiously claustrophobic. It may seem churlish to complain about the instrumentation and switchgear, but there is no denying that neither is particularly ergonomic or at all well labelled; only the speedometer and rev counter are really convenient, but then again, they are what you need most!

Going the other way, there was also a bigger derivative of the Lampredi engine for 1956, the 410 Superamerica. This was equipped with a 4,962cc motor delivering anything from 340 (Series I) to 400 bhp (Series III) and propelling a car that weighed perhaps 2,420 lb. (1,100 kg), which meant that the power-to-weight ratio was between 310 bhp/tonne and 364 bhp/tonne. Only 38 of these beasts were built, but top speeds of over 160 mph (260-270 kph) were possible with the proper rear axle ratio. The Series III (the most powerful model, introduced in 1958) would cost anything upwards of $16,000, *in Italy*. The very last Lampredi-engined model, the Type 400 Superamerica, claimed 400 bhp at its introduction in 1960, but later the claimed power output dropped to 340 bhp. Rather over 50 400 Superamericas were built, but they were not 4.8 liter cars as one might expect; rather, the "400" referred to the total capacity in deciliters, as the actual capacity was 3,967cc.

Thus far, Ferrari had stuck to the front engine/rear drive approach with which they had begun. By the early 1960s, though, it became increasingly apparent that mid-engined cars were the wave of the future as far as state of the art sports cars were concerned, and so Ferrari stuffed their long-suffering V12 into the middle of a new car, the Type 250LM.

Vintage Wine in New Bottles

The Colombo engine was mounted longitudinally in front of the rear transaxle for the 250LM, resulting in a driving position that was well towards the front of the car. The LM stood for Le Mans, and the whole cocktail was very exotic indeed.

It was back to the old days of limited production, with only 35 to 40 of the cars produced, and despite the "250" designation, the majority of them actually had 3,286cc motors delivering 320-330 bhp; the original three-liter motors delivered about 300 bhp. Given that the cars weighed rather under 1,900 lb. (a fraction over 850 kg), the power-to-weight ratio was a dramatic 353 bhp/tonne to 388 bhp/tonne, and these things could just) be driven on the road. Top speed was in excess of 170 mph (275 kph) with the right gearing, and the handling was superb. Instead of the old-fashioned chassis, the frame was a combination of triangulated tubes assisted by the aluminium skin as a stressed part of the structure: wherever access was not required, the skin was pop-riveted on and served to create a monocoque/space frame chassis. Four-wheel disks were fitted (inboard at the rear), and the whole thing was about as different from the by now very civilized 250GT as it could be.

There were, however, certain compromises that one had to make. To begin with, it was a very hot car to drive. Not just in the slang sense, but literally. With the engine immediately behind the driver, the radiator and oil cooler were in front, and the pipes that connected the two warmed the cockpit liberally. Luggage space was effectively non-existent. On the other hand, it was remarkably similar to the car that won outright at Le Mans in 1964 ....

Although the 250LM appeared in March 1963, a much more traditional Ferrari was to appear less than a year later. It was a bigger-engined replacement for the old 250GT, with 3,967cc and 300 bhp. By Ferrari standards, though, it was remarkably portly at 3,040 lb. (1,380 kg), and not everyone loved it – though it certainly sold well enough, with a total run of about 1,000 cars over its long model life.

There was also a whole new V12, generally regarded as a Colombo derivative though also having some Lampredi characteristics, in the Type 500 Superfast. Again, the "500" referred to the overall capacity (4,962cc) and the motor delivered 400 bhp. The car was a derivative of the 400 Superamerica, and at 3,200 lb.,1,455 kg, it was very heavy indeed for a Ferrari. Even so, it could still boast 275 bhp/tonne, and top 170 mph (274 kph). Only about 37 were built before the model was discontinued in 1967.

In October 1964, though, another "baby" Ferrari had appeared. It was known as the 275GTB ("Berlinetta" coupé for racing) and 275GTS ("Spyder" convertible for touring). Actually, it was a mixture of old and new: still the inevitable Colombo-derived engine in 3.3 liter guise, with 260 bhp (GTS) or 280 bhp (GTB), and still a front engine/rear drive car, but much else had changed.

The chassis was now a multiple-tube affair instead of the old ladder, and the gearbox was incorporated in the

9

rear transaxle. Suspension was independent at all corners, and the steel-and-alloy or all-alloy body was built by Scaglietti to a Pininfarina design. Weighing only 2,550 lb./1,160 kg in coupé (racing) form, or 2,750 lb./1,250 kg in convertible "Spyder" form, power to weight ratios were 241 bhp/tonne and 208 bhp/tonne respectively. If 100 bhp/tonne is the base-line for sporting cars, as it is often held to be, then over 200 bhp/tonne in road-going trim is something quite remarkable.

For 1966, the old Colombo V12 finally acquired four cams; this was the 275GTB/4, which had 300 bhp and which was usually installed in stripped-out bodies weighing as little as 2,400 lb. (1,100 kg). The 330 engine was also made available in the 275 body, as the 330GTC and 330GTS; about 600 330 GTCs and 100 330 GTS convertibles were built.

In due course, the 330GT grew to a 365GT: about 3,500 lb., 1,600 kg, which even with 330 bhp from the 4,390cc long-stroke derivative of the Colombo engine was only just over the 200 bhp/tonne figure which was becoming *de rigueur* for Ferraris. This was, however, a remarkably luxurious 2+2 which was also very quick at anything up to 150 mph, over 240 kph.

Then, perhaps predictably, the 365GT got some extra cams to power the 365GTB/4 "Daytona," a derivative of the 275GTB/4 with 352 bhp — which it needed, given that the car commonly weighed 3,600 lb. or about 1,635 kg. The "Daytona" was Ferrari's most expensive car, but it sold like hot cakes between 1968 and 1974; over 1,400 were built, including 127 Spyder convertibles. The top speed of the coupé was a claimed (and sometimes verified) 175 mph, about 280 kph. A rather lighter GTC/4 (about 3,190 lb./1,450 kg) was made in 1972-73, powered by a 320-340 bhp engine. All sorts of bodies were available, including even four-seaters.

The front engine/rear drive Ferrari V12 cars had, however, pretty much reached the peak of their development some time before the early 1970s; as we shall see below, much more exotic cars were now on the way in, so the V12 story need only be rounded off. The Type 400i was the last really new V12, and it was a four-seater family car — albeit a family car as only Ferrari would do it. The 1976 Paris show saw a fuel-injected, 4,823cc version of the V12 delivering 340 bhp but weighing a staggering 4,000 lb. (1,850 kg or so). Intended as a replacement for the former 365GT4 2+2, it was a splendid car but was really something of a fossil. It finally disappeared in 1989, but in 1992 a "new" Ferrari appeared in the same mold, the 456. At the time of writing, this car had just been launched at the Paris show, and it was by no means certain what changes might be made for the production vehicle, but it was essentially a 5.5 liter 2+2 with a V12 at the front and rear wheel drive. The projected price was $200,000 upwards.

Dino

At the 1965 Paris Salon, an all-new Dino 206S Speciale Coupé appeared, with a mocked-up two-liter, four-cam V6 mounted transversely behind the cockpit and driving the rear wheels. The car eventually entered production in 1969, though without any Ferrari insignia; the "Dino" name was a tribute to the son of the *Commendatore*, who had died young. Although not strictly Ferraris, Dinos are normally included in the Ferrari canon, and indeed in 1976 the familiar prancing horse finally appeared on what had hitherto been called a Dino.

The original show-car chassis was a racer, and the production chassis was derived from it. The body design was by Pininfarina, though the actual construction was by Scaglietti. The engine, with its 65 degree included angle, was actually built by Fiat. In its original two-liter form, the all-alloy engine delivered 180 DIN bhp, which in a car that weighed 1,980 lb. (900 kg) gave a highly satisfactory 200 bhp/tonne. Triple Weber twin-choke downdraft carburetors kept the motor supplied with fuel; obviously, tuning the V6 was very much easier than tuning the V12, where trying to synchronize twice as many carburetors can be a very time consuming process. If the carburetors are not properly synchronized, however, the penalty is horrific fuel consumption, especially on the V12.

Top speed was 142 mph (say 230 kph), and the 0-60 mph time was just over seven seconds, but the wonderful thing about the Dino was its balance. All kinds of nonsense has been written about the Dino, because it was "only" a V6, and "only" 1,987cc, and could "only" achieve 142 mph, but as a machine to be driven very quickly indeed, the Dino is surely a very desirable car. By Ferrari standards it was also a mass-production car; about 4,000 were made between 1967 and 1973.

After the first 100 or so cars, it became even more desirable when the engine was enlarged to 2,418cc in 1969; the Dino then became known as the 246. A

retrograde step was the adoption of cast iron for the block, though the nominal weight remained the same and power was increased to 195 bhp. For 1972, a "Targa" style removable top was offered (the 246 GTS), but by this time even the fixed-head Dino had put on a considerable amount of weight and was up to 2,380 lb. (1,082 kg). This represented a weight increase of some 20 per cent, which with the same 195 bhp necessarily implied a reduced power-to-weight ratio: "only" 180 bhp/tonne.

In order to remedy this, the Dino was completely redesigned, and in 1973 it appeared with a new Bertone body and a new three-liter V8 engine: the 308GT4.

It was an odd-looking car, with an enormously long mid-section and a short, sloping nose, but the extraordinary thing was that it was not a two-seater (and a close fit, at that), but a 2+2. Some even called it a four-seater, though the driver and front passenger would need to be fairly short-legged if the rear passengers were to have sufficient knee-room for anything but a short journey, and it was by no means unusual for the car to be supplied (at the customer's request) with additional luggage space in place of the rather nominal rear seats.

The V8 engine, unlike the Fiat-built V6, was an all-Ferrari unit with a 90 degree included angle. It delivered 255 DIN bhp in European trim, with somewhat less power (sources are in conflict) for the American market. It was virtually guaranteed to shock the purists: everyone knew that "real" cars (and especially "real" Italian cars) shouldn't have V8 motors, which were mere Americana; somehow, the fact was conveniently overlooked that Rolls Royce and a number of other eminently "real" cars, including several from the United States, came with V8 power. Worse than the adoption of the V8 layout, though, was the way that the four cams were driven by toothed belts instead of chains. Never mind that camshaft chain tensioning on the forward bank of the V6 was a well-known problem (retensioning them was inconvenient, and was therefore sometimes omitted by mechanics): the mere idea of belt-driven cams was the purest heresy. To confuse matters further, there was a two-liter version of the same V8 for the Italian domestic market, because of Italian tax laws: cars over two liters were taxed at 35 per cent, while cars under two liters attracted only 20 per cent, so even to a Ferrari buyer the difference was significant. The 308 with the smaller motor was predictably called the 208.

Regardless of all this, as installed in the 308GTV the V8 was capable of propelling 2,930 lb. (1,332 kg) of motor car at up to about 150 mph (over 240 kph), even if the power-to-weight ratio had fallen to just over 190 bhp/tonne. According to whom you believe, U.S. customers had to get by with either 205 or 240 bhp, for either 154 bhp/tonne or 180 bhp/tonne. This may not meet the rarefied demands of some Ferrari *aficionados*, but it is still not too bad! Also, the adoption of the V8 format was a pointer for subsequent Ferrari models. The Dino itself began to sport prancing-horse insignia on the nose, wheels and steering wheel in late 1976, though no official explanation was given for the change, and it remained as a Ferrari until it was discontinued in 1979.

One other variant, which appeared only under the Ferrari name, was the 208GTBi (and GTSi) Turbo, another only-for-Italy car, with 211 bhp and a claimed top speed of just under 150 mph, about 240 kph. This was Ferrari's first road-going turbocharged car; their main problems were, in the first place, getting the turbocharger into the engine bay (there's not much room in a mid-engined car), and in the second place keeping it cool, which they achieved with extra NACA ducts which actually enhanced the appearance of an already beautiful car.

The V8 Generation

The somewhat portly 308 2+2 was obviously not the best place to show off the new V8, and at the 1975 Paris Motor Show the new two-seater 308 GTB appeared. This time, the body reverted to the Pininfarina/Scaglietti formula, though (somewhat improbably) the first run of bodies – maybe 300 – made extensive use of glass-reinforced plastic (GRP).

The 308GTB was a far less practical car than the 308GT4, with minimal luggage space. What was worse, the main trunk was just behind the engine and got very warm indeed: excellent for clothes, but not much good for champagne or bars of chocolate. Visibility, except to the front, was very poor, and the car actually weighed more than the 2+2: 3,160 lb. (1,436 kg) even as a berlinetta, and no less than 3,225 lb. (1,466 kg) in the Targa-top GTS form which was introduced in 1977 at Frankfurt.

It was, however, a very much better looking car than the 308GT4, and it was aimed squarely at the U.S. market, with great emphasis on creature comforts. Also, it seems to have been the first Ferrari ever designed with

11

a long production run in mind: it was not replaced until 1990, a fifteen-year production run in which some 3,665 examples were built.

The original engine was the same as that in the Dino GT4, with 255 bhp for the U.S. market and 255 bhp for the rest of the world, and the 208 engine continued to be available in Italy, though for 1981 the small engine was turbocharged for the Italian market. In the same year, the bigger engine switched from carburetors to Bosch K-Jetronic fuel injection. Power remained much the same, but once the engine was set up, it stayed set up, and fuel economy and tractability were much improved. Then, in 1983, a *quattrovalvole* (four-valve) head was added, for about a 10 per cent increase in power. In 1985, a modest increase in capacity to 3,185cc brought still more power, so that even the Americans now had 260 bhp.

When at last the 308GTSi and 308GTBi were replaced in 1989 (though there was some overlap, with the older cars remaining available), it was with a similar-looking car which was, however, rather more different than it seemed at first sight. The V8 had now been enlarged to 3,405cc with up to 300 bhp, but it sat five inches (125 mm) lower in the chassis. The wheelbase was four inches (100 mm) greater, with minimal rear overhang, and the whole car was significantly wider. The weight remained much the same, though, at 3,170 lb. (1,441 kg), so the power-to-weight ratio climbed back once more over the magic 200 bhp/tonne figure.

Meanwhile, the V8 had also been installed in yet another car, a grand tourer called the Mondial ("Worldwide"). First shown at Geneva in March 1980, the Mondial was clearly a *de luxe* Ferrari, with Connolly leather seats, air conditioning, central door locking electric mirrors and the other gewgaws which were becoming *de rigueur* for the American market. The body was designed by Pinin-farina, and clothed a mid-engined chassis of remarkable performance. The only trouble was, the car was far too heavy for a Ferrari, so the full capability of the chassis was never realized. Also, because it was perceived as a slow Ferrari, buyers stayed away in droves.

There were various updates, modifications and engine improvements, but the turning point came in 1983, when the Mondial was made available as a true convertible, not merely a "Targa." Very handsome it was, too, but the weight of the fixed-head car was 3,400 lb. (1,545 kg) then, and by 1990 the weight of the convertible had crept up to 3,462 lb. (1,574 kg).

The transversely-mounted engine marched in step with the 308 series, going first to a 32-valve layout and then acquiring extra capacity; the Mondial t was the 3,405cc version, introduced in 1989, with a revised engine position and a stretched wheelbase. By any normal standards, a Mondial is a fast car – but fast is a relative term.

This is well illustrated by comparing it with the GTO of 1984, which was literally *Omologato*, a homologated FISA Group B racer of which 200 examples had to be built in order to qualify for homologation. As fitted to this 2,555 lb. (1,161 kg) road rocket, the V8 was equipped with twin IHI turbochargers and delivered 400 bhp. In this form, the engine had actually been developed for Lancia rally cars (both Lancia and Ferrari were under the Fiat umbrella), but it made for a truly spectacular road-going car with 345 bhp/tonne and a top speed in excess of 186 mph (300 kph). Zero-to-sixty times were about five seconds. The engine was mounted longitudinally, pointing the way for later mid-engined Ferraris, including eventually the Mondial.

At the time of writing, the latest Ferrari V8 was a somewhat lateral derivative of the above. The F40, introduced in 1987 to commemorate 40 years of Ferrari road and racing cars, used a 90 degree V8, but with a larger bore and shorter stroke than even the original V8: 3.23 x 2.74 inch (82 x 69.5 mm), as against 2.19 x 2.8 inch (81 x 71 mm), while the 3,405cc motor was 3.35 x 2.95 inch (85 x 75 mm). From a swept volume of 2,936cc, twin IHI turbochargers helped the engineers to extract over 470 bhp: reported figures, and no doubt individual power outputs, varied. Despite extensive use of exotic materials, including Kevlar and carbon-fiber reinforcements, and despite a very basic and stripped-out interior, the F40 weighed in at 2,425 lb. (1,102 kg), but the aerodynamics provided enormous down-force, an essential characteristic in a motor-car which could easily exceed 186 mph (300 kph) and where 200 mph (323 kph) was tantalizingly close. In the United States, anyone who wanted an F40 would have to find something like $400,000 – and for the first few months there was a waiting list. Then the world economy turned sour, and suddenly the F40 was readily available – and you could get one for under $300,000. It remained in production for about three years.

Incredibly, a *Los Angeles Times* road test of the car drew attention to the lack of creature comforts, commenting

on the absence of a radio (which you would never be able to hear properly anyway) and remarked on the pull-wire door closing arrangements – and even more incredibly, Ferrari eventually responded by adding a more conventional interior, even at a considerable weight penalty, though it was still pretty Spartan.

Not only were the windows not powered, they were not even wind-up, and for that matter, they were not even glass. Sliding plastic side windows save weight and are perfectly adequate for fast driving! Also, vision to the sides and rear was appalling: this was not a car for driving in traffic. No power steering, no anti-lock brakes, very little in the way of interior sound insulation – the list of amenities that the F40 does not have is considerable. But how many other street-legal motor cars can you compare with a Ferrari F40?

Meanwhile, in addition to the V12, V8 and V6 layouts, Ferrari had also tried another layout: the flat-12.

Boxers

The boxer, or horizontally opposed motor, has several striking advantages, not least an excellent primary balance which allows the engine to rev faster and more smoothly than a "V" configuration of similar displacement and number of cylinders. The flat twin is always associated with BMW motorcycles; the flat four with Volkswagens; the flat six with Porsches and, as far as roadgoing cars are concerned, the flat twelve is a Ferrari preserve. There have been other flat twelves, but they have all been exclusively racing engines.

Ferrari's very first flat-twelve was in fact a 1.5 liter Formula 1 engine, which appeared in 1964, but the first road car with the flat-twelve engine appeared at the Turin motor show in October 1971. It actually went into production in 1974 as the 365 GT4 BB (365 Grand Touring 4-cam Berlinetta Boxer), with the cams driven by toothed belts. The motor was mounted longitudinally between the driver and the rear wheels, and was clad in a very handsome two-seater body designed by Pininfarina and built by Scaglietti. The main structure was steel, but the lower body panels (which were always painted black) were of GRP, and light alloy was used for the front hood, the engine cover, and the doors.

Instead of the traditional oval-tube and round-tube construction, the chassis was square and rectangular tubes, and the 365GT4BB was a big, heavy car at 3,420 lb. (1,555 kg). On the other hand, it was also a big, powerful engine: 4,390cc and 344 DIN bhp, for 221 bhp/tonne. The mixture was supplied by four triple-choke Webers. Although it lacked the nimbleness of a smaller, lighter car, the rearward weight bias (43/57 per cent) was to some extent "faked out" by the steering and suspension geometry. The top speed was an impressive 175 mph (over 280 kph).

For 1977 the Boxer motor was increased in capacity to 4,942cc by the expedient of an extra millimeter of bore and seven millimeters more stroke. This brought the power up to 360 bhp, and the top speed up to 188 mph, which is just over 186 mph – the magic 300 kph. Despite the enormous power and very high top speed, the car was reviewed as being remarkably tractable and forgiving, though one U.S. publication complained that the gearing was too "tall" – which it probably was, if all you were interested in was drag racing. With the five-liter engine and minor body modifications, including a small "chin" spoiler, the new Boxer was known as a 512 BB. Remarkably, when the change was made from the Webers to Bosch K-Jetronic fuel injection in 1981, the stated horsepower actually dropped to 340. In this form, the 512BB continued into 1985.

If, as *Road and Track* had stated, the 512BB was the best all-round sports and GT car they had ever tested, they must have run out of superlatives when they tried its replacement, the Testarossa.

This was first seen at the 1984 Paris auto show, though (as usual with Ferrari) there was a fair lag between introduction and series production. The name, which means "Redhead," was borrowed from a Ferrari sports-racer of the late 1950s, which was so called after its red cam boxes. The engine was a further development of the flat-twelve, with four valves per cylinder (48 valves in all!) and 380 bhp under the SAE nett rating; maybe 390 bhp or even 400 bhp DIN.

The car was slightly lighter than the original 365 GT4 BB, though 44 lb. (20 kg) is not really all that significant in the context of more than a tonne and a half of automobile. On the other hand, the new Pininfarina-designed and Pininfarina-built body was startlingly beautiful (though there were those who said that it was simply startling). As dramatic as the wedge shape were the huge air intakes behind the doors, which were needed to cool the rear-mounted oil and water radiators. Although moving these to the back did nothing for front/rear weight distribution,

13

it did everything for cockpit comfort: previously, the hot oil and water had been routed to radiators in the nose, and warmed the driver and passenger rather too well. The "cheese slicer" strakes which began on the door and ran back to the air intakes were necessary because the construction and use regulations in some countries required that holes above a certain size had to be covered with a grille of some kind. The long intake trough was found to be necessary as a result of wind-tunnel testing.

Further insulating the occupants from the stressful effects of the flat-twelve was a brilliant and original solution to the problem of carrying luggage in a mid-engined car: special fitted luggage went into the space behind the seats. The passenger compartment was also as luxurious as could reasonably be expected; the Testarossa was designed to be a practical car (or at least, as practical as any car of that price and that speed is likely to be), and all due attention was paid to creature comforts.

Moving the radiators to the rear also allowed for more luggage space up front, and when all was said and done the Testarossa genuinely was a practical and very fast grand tourer. No one seems entirely sure what the top speed might be: some claimed over 186 mph (300+ kph) as the 512BB, while others estimated a top speed as low as 175 mph (282 kph). Such figures are, however, doubly meaningless. First, there are very few places in the world where such speeds can be attained on public roads, and still fewer where they can be attained legally. Second, maximum speed figures for ultra-high-performance cars frequently involve taking the rev counter deep into the red zone, with attendant risks to the engine. While this is all very well for the factory and for motoring journalists, most owners are likely to be slightly more circumspect with their own cars. Even if they can afford the expense of a rebuild (and one assumes that most of them can), there is still the inconvenience to consider.

Anyway, it didn't matter. Unless you were a very highly skilled racing driver, you would need either to be certifiably insane or very, very careless indeed in order to run a Testarossa outside its "handling envelope." With a price approaching $200,000 at the time of writing, you would also need to be remarkably rich.

Racers, Fours and Sixes

From the very beginning, Enzo Ferrari's main interest was in racing. The original 1,500cc Colombo-designed *Tipo* 125. was designed to be supercharged as a Formula 1 motor, under the rules then obtaining about a three-to-one differential between supercharged engines and those with normal aspiration. Then, the Lampredi "long block" was designed as a conventionally-aspirated Formula 1 motor: 4,500cc is, of course, three times as big as 1,500cc.

To Ferrari's chagrin, the V12 never delivered as much power in either form as he had hoped: the *cansone del dodeci*, the "song of the twelve," turned out to be a siren song.

For serious racing, Lampredi therefore turned to the other extreme. He believed that the V12 delivered less power than it should because it was too heavy, too complex, and too burdened with frictional losses; as a matter of interest, current theoretical studies suggest that a V10 is the ideal compromise between frequency of firing pulses and frictional losses.

Be that as it may, Lampredi's Type 500 was a 1,985cc motor with four 500cc cylinders, using a 3.54 inch (90 mm) bore and a 2.76 inch (70 mm) stroke: strictly, it should have been a Type 496. The same unusual cylinder arrangement was used as in the Lampredi V12, with the liners screwed into the head, and twin overhead cams actuated the valves. In the hands of Alberto Ascari, it won the World Championship in both 1952 and 1953, and gave rise to the 500 Mondial.

This was a sports/racing Ferrari with two twin-choke Weber 42DCOA3 carburetors – effectively one carburetor per pot – and a 9.2:1 compression ratio (Formula 2 cars were 13:1). The heads were twin-plugged to allow adequately fast flame spread. Early versions delivered 170 bhp, later ones 185 bhp, and in a car that weighed only 1,715 lb (780 kg) this meant 220 bhp/tonne. The passenger seat of the Mondial was normally covered with a metal tonneau, and the spare wheel took up the whole of the "boot," but it could still be driven on the road, and it was.

The 750 Monza was a three-liter derivative of the Mondial, with a lower compression ratio (about 8.6:1) which allowed it to run on pump fuel. Giant twin-choke Webers fed the beast, much like the Mondial, and power output was about 260 bhp. It weighed about the same as the Mondial, and was a pig to drive; it was in a Monza that Ascari died, after borrowing the car to see what it was like.

Other four-cylinder derivatives included the 553 F2 (two liters); the 625 (2.5 liters); the 735 (three liters); the 857S (3.4 liters); and the 860 (almost 3.5 liters). The

most famous, though, was the 500TR "Testa Rossa" or "Red Head" of 1955; the name came from the cam-boxes, which were painted bright red. This was one of the most successful four-cylinder racers of all time, and was also one of the last racing fours ever to be really successful.

Also in 1955, the four was "stretched" to make a straight-six: there were two models, the 376S or 118 (3.7 liter six, factory project 119) and the 121 or 446S or 735LM (factory project 121, 4.4 liter six, 735cc per cylinder Le Mans). Perhaps inevitably, the sixes won some events including the Tour of Sicily, but the fours were more successful and the sixes lasted only for a single season.

None of these fours should be confused with the "Ferrarina" or "baby Ferrari" which seems to have existed in 850cc and 975cc form: the factory prototype bears no Ferrari badges, but the double overhead cam engines delivered 80 bhp (850) and 86 bhp (975); the "854" cast into the cam box seems to refer to "850cc, four cylinders." Although it never went into production as a Ferrari, the design was sold to the De Nora family, who planned to build a car – the ASA Mille – to compete with the Porsche. A few were built, but they could not compete, so the project died.

After the early-1950s fours and sixes, the Formula 1 and Formula 2 Ferraris diverged ever more widely from the road cars and from the sports/racers, which tended to be more like the road vehicles. A full range of engine layouts appeared, including flat and V configurations, from six to twelve cylinders, but the countless variations even in the road-racing cars, let alone the Formula One cars, would require a book to itself.

Memories and Dreams

Enzo Ferrari died on the August 14, 1988. He had lived long enough to see the F40, the fastest road Ferrari ever built. The name commemorated forty years of Ferraris, from the first racing season of 1947/48 to the last racing season that the *Commendatore* would ever see.

In those forty years Ferrari had gone from making essentially hand-built cars to modest series production: from three cars in 1947, and five in 1948, to four thousand cars a year by the late 1980s. In fact, more Ferraris were made in 1987, the year before Enzo's death, than were made altogether from 1947 to 1964 inclusive.

The number of employees had risen from about forty to over seventeen hundred.

The years had been stormy. The death of Enzo's son Alfredo in 1957 had been a great blow to the *Commendatore*, who was approaching sixty, and by 1965 he was ready to consider selling out to Ford. Fortunately, he did not: from all that one can learn, it seems that Ford wanted the name and a racing *scuderia*, but had little idea of how Enzo Ferrari ran the show. They apparently could not believe that he expected to continue to run things his way after he had sold them the company, and Enzo Ferrari would never have survived as the kind of 1960s Company Man who was then the Ford corporate ideal. Eventually, in 1969, Ferrari effectively sold out to Fiat – but they were prepared to let him do things his way and, apart from a massive injection of capital, Ferrari continued much as it had always done. By 1965 Ferrari production was running at around 700 cars a year, and the 1,000-a-year barrier first fell in 1971. It would not be until 1979 – a full decade after Fiat's involvement – that production exceeded 2,000 a year, and it would be almost a decade later again before 4,000 cars a year were made.

All of these numbers may seem very dry, but they tell the story of Ferrari with remarkable clarity. The history of Ferrari falls into three sections: very roughly, the first fifteen years, then the next quarter-century, and finally the brief period since Ferrari's death, when the future of the company has somewhat hung in the balance.

The early days were marked by the full fury of Enzo's dictatorial ways. The road cars were barely-modified racers, and were sold essentially to finance the racing program. The cars which are now regarded as the early classics are (and always were) fabulously rare, and most Ferraris from the 1960s are not much more common. Few were built in the first place, and of those few, not all have survived. The tin-worm has claimed many: as we have seen, Ferraris were never very strong on rust-proofing. Others have been crashed and written off, some "legitimately," in races, and others because they were bought by people who were unable to drive them properly.

The second period saw the beginnings of standardized production-line Ferraris. It began before Fiat was involved, and by Ferrari's death was an accomplished fact. During that period, the cars increasingly fell into three separate camps. There were the "cooking" and "family" Ferraris, like the 400, the Mondial and most of the other V8 cars;

15

there were the fast luxury Ferraris, like the flat-twelves; and there were the seriously fast Ferraris, like the two series of GTO and the F40, where few if any concessions were made to driver comfort. These were the true heirs to the old Ferrari concept of the racer on the road.

What remains to be seen is what will happen now. Clearly, design projects have a life of several years, and we are only now seeing the last of the cars which were new ideas when Ferrari was an old man. The F40 is out of production, leaving no heir to the line of what I have called seriously fast Ferraris – but then, there have been other times without racer-on-the-road Ferraris.

It may be significant, though, that the most recent "new" Ferrari was a very traditional car, the 456. A front-engine, rear-drive V12 with 2+2 seating capacity, it was actually the least "new" Ferrari for a very long time, but it was the kind of Ferrari that attracts the wealthy, middle-of-the-road buyer who wants a sleek, luxurious, fast but still relatively practical car. The prancing horse badge and the V12 motor (instead of the more pedestrian V8, which simply lacks the *cachet* of a twelve) were guaranteed to attract the same kind of market as Aston Martin/Lagonda, or possibly Bristol, or those who remembered Iso and Pegaso fondly – and, unlike the F40, the price was competitive with Aston Martins as well. To be brutal, the 456 was simply a more exclusive (and less practical) version of a Jaguar XJ-S, another car which symbolizes the "end of history" approach to car design.

What do I mean by "end of history"? I mean a car which already does just about everything that can be expected of it, and which cannot really be much improved upon. It cannot be made much more luxurious, without degenerating into decadence. It cannot be made much faster, because there is nowhere to drive a car that is much faster, and there are in any case few drivers who could handle more speed and more acceleration. And, perhaps most significantly of all, it cannot be made much sweeter-handling without significant sacrifices in comfort, because comfort means weight, and a car that weighs much less is simply not going to be acceptable to enough people, with enough money, to sell in adequate numbers given Ferrari's 4,000-a-year production capacity. Ferrari is not big enough on the one hand to build "flagships" like the Jaguar, and not small enough on the other hand to build race-shop specials like the McLaren.

Furthermore, it has been some time since Ferrari built the ultimate supercar. Even the F40 has been upstaged by a number of others: the Jaguar 220, the born-again Bugatti, the revised Lamborghini Diablo, the McLaren F1. Whether or not these vehicles actually are superior to the F40 is irrelevant, because only a handful of people are ever going to be able to afford them, but the important thing is that the Ferrari F40 was seen by Ferrari buyers as an irrelevance, and by schoolboys and the motoring press as second-best.

This may be a completely wrong reading of Ferrari's future. Even if it is correct, it is far from a condemnation of Ferrari. After all, the pre-Derby Bentleys were the Ferraris of their day: handsome, desirable, and wickedly fast. Likewise, Aston Martins are no longer perceived as cutting-edge racers. This does not stop people wanting to buy either Bentleys or Aston Martins. And again, even if it is correct, it does not mean that there will cease to be ultra-fast "flagship" Ferraris from time to time. But to sum up, I would not mind making a small bet that the F40 and even the Testarossa represent a strain of Ferrari which is going to become less and less important in the future, while "practical" Ferraris such as the Mondial t and the 456 continue to be the bread and butter of the company from Maranello.

1950 FERRARI 166 MM BARCHETTA

The earliest Ferraris, like this 1950 166MM "Barchetta" by Touring, are (and always were) fabulously rare. To the modern eye, there is incredibly little under the hood – just a single carburetor on the V12 engine, for example – and while the body may be beautiful, there is not a spare ounce of weight on it. In the bare cockpit, there is no lining for the doors and there is no provision for weather protection. The Veglia "clocks" run counterclockwise; as on any racer, the tachometer is of course nearest the driver.

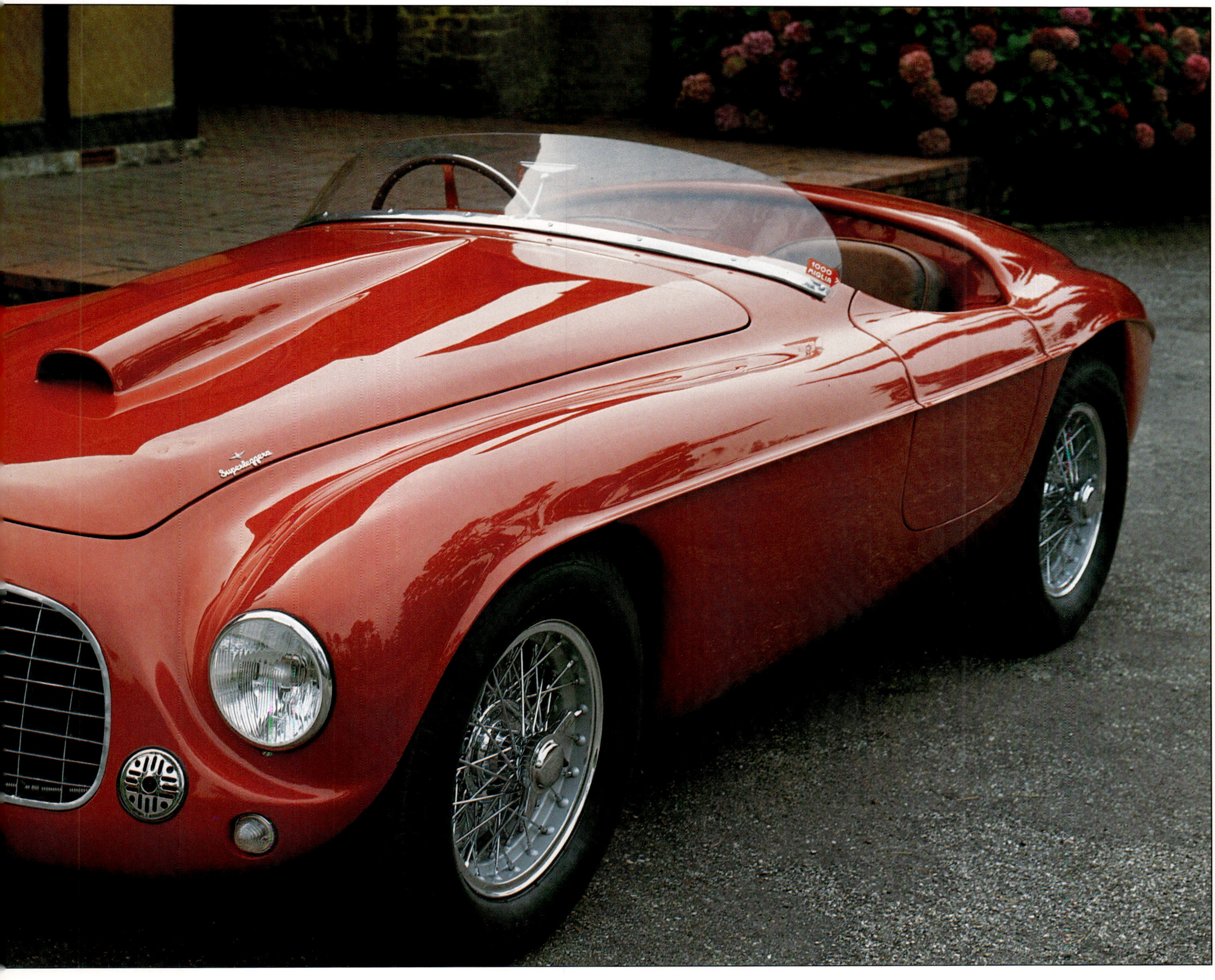
Superleggera
1000 MIGLIA

1952 FERRARI 340 MEXICO

The 340 "Mexico" took its name from Ferrari's success in the 1951 Carrera Panamericana, one of the most grueling of road races, which ran right across Mexico. The huge flaring front wings are almost a parody of styling, and were aerodynamically far from efficient. Vignale bodied just three coupes in this style.

21

1952 FERRARI 625 TF

The 625TF is something of a mystery: a late 166-type chassis with Vignale bodies like the Mille Miglia cars, it was powered by a twin-plugged, twin-cam four-cylinder engine apparently identical to those used in Grand Prix cars. The "TF" name comes from "Targa Florio," for which the cars were apparently prepared; the vehicle is also known as a 625S. The first competition appearance of the 625TF was at Monza on June 28, 1953, driven by Mike Hawthorn. The car is liberally supplied with cooling scoops and extractors, but the overall effect is still a work of art, rather than being a mere mechanical contrivance.

1953 FERRARI 375 AMERICA

24

The 375 America is a classic example of the way in which modern luxury Ferraris derived from the original light, Spartan racers. The 375MM (Mille Miglia), on which this car was based, was a no-compromise racer, but the same chassis made a gorgeous road car when clad in a body by "Pinin" Farina; this was before he ran his nickname and his given name together to become Pininfarina. The engine was downrated somewhat – by a little over ten percent – when compared with the racer.

1954 FERRARI 250 GT EUROPA

The history of the 250 Europa is complex. The earliest Europas used the Lampredi engine, but the return to the Colombo design in three-liter guise marked a turning point in Ferrari's history. Only about 21 Lampredi Europas were built, but the Colombo Europa (later the 250GT) became Ferrari's first "production" car.

1956 FERRARI 410 SUPERAMERICA

"America" was the name that Ferrari used for his larger-engined cars, and the word "Superamerica" was coined to describe his most luxurious car to date, the 410 Superamerica. Built with a clear eye to the American market from 1956 to 1959, this car (also seen overleaf) shows some American styling influences as well as an un-Ferrari-like regard for creature comforts.

1957 FERRARI 250 GT CALIFORNIA

The 250GT California was very much the brainchild of Luigi Chinetti, longtime Ferrari enthusiast and racer, and Ferrari's importer into the United States. The looks are a blend of American and Italian, but the performance is pure Ferrari.

1958 FERRARI 250 GT ELLENA

The first 250GT cars (also overleaf) were bodied by Boano/Ellena, and were surprisingly dignified and restrained. The front is reminiscent of Farina, and the rear of Michelotti, but the whole package was elegantly put together and made for the most practical Ferrari yet.

35

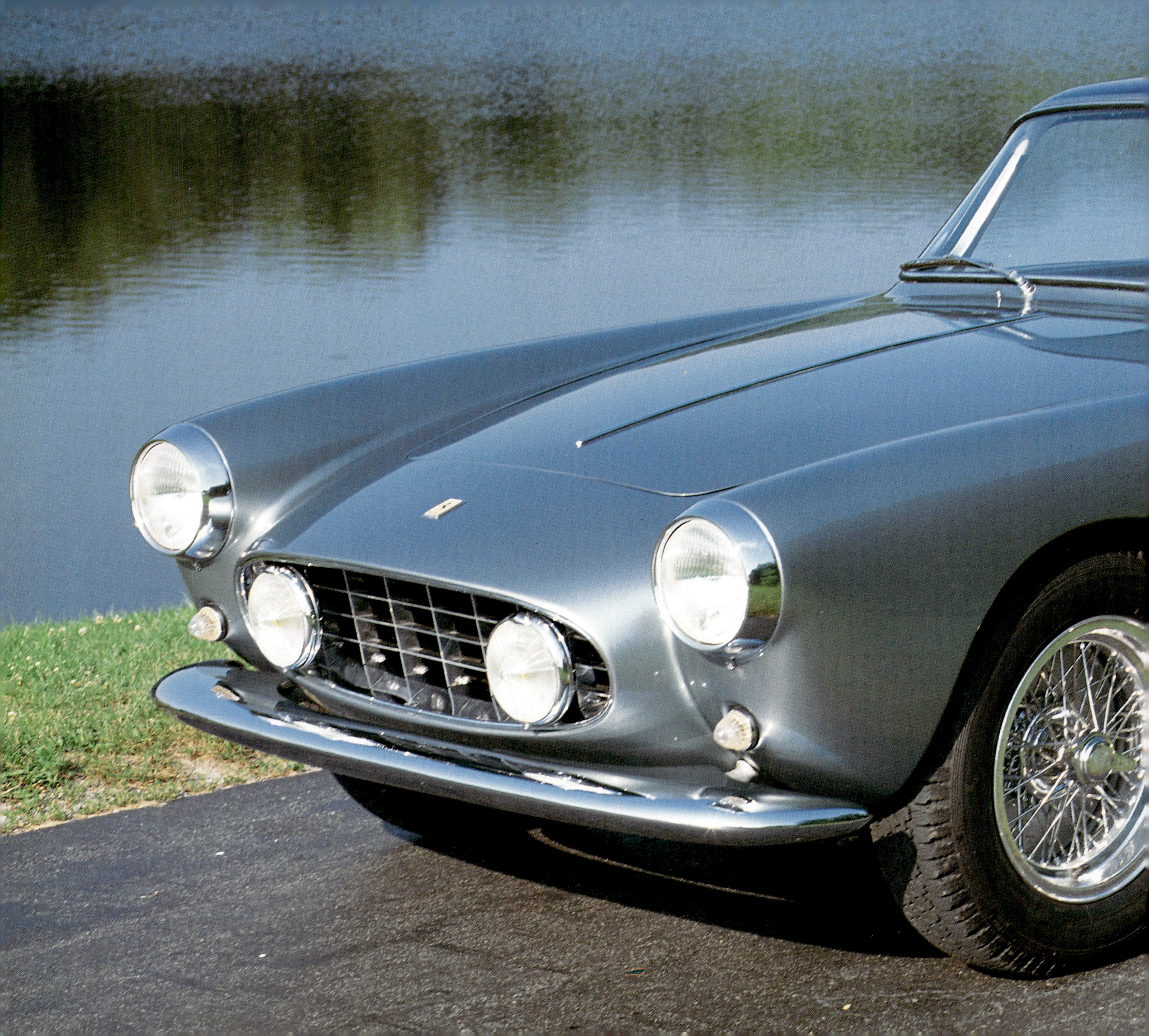

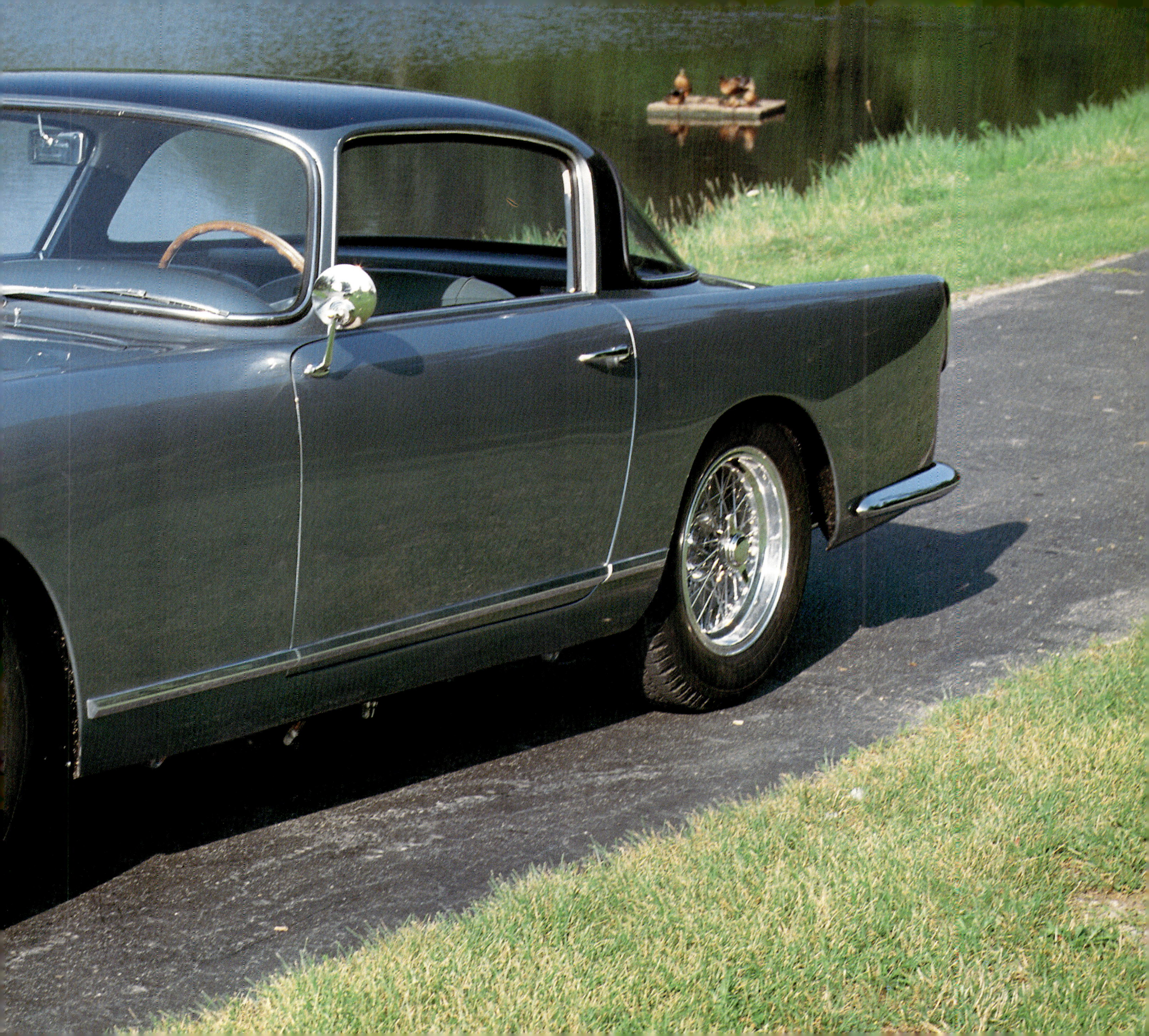

1958 FERRARI 250 TESTA ROSSA

The original Testa Rossa was a four-cylinder car, but the 250TR V12 Testa Rossa (also seen overleaf) became rather better known. The cutaway "pontoon" wings were supposed to aid brake cooling, but were aerodynamically unsuccessful. It was still a very traditional vehicle, with front engine in unit with the four-speed gearbox, and a live rear axle suspended on leaf springs – arguably, the peak of development of the vintage racer.

1958 FERRARI 250 TESTA ROSSA

41

1958 FERRARI 250 CALIFORNIA

To modern eyes there is perhaps a little more chrome on this 250GT California Spyder than is entirely appropriate; but you have to look at this car through the eyes of the 1950s, not the 1990s, comparing it with, say, a Corvette. For the time, it was wonderfully restrained, and a superb combination of elegance and understated confidence. The styling was by Pininfarina and actual construction was by Scaglietti. The fully instrumented interior was a sports car lover's delight, as well as being remarkably comfortable. This is a Series 2 car, produced in late 1958; twenty-seven were built before the Series 3 came out in 1959. This rarity, together with the glamor associated with both Ferrari and California, makes it a very desirable and valuable car today.

1958 FERRARI 250 CALIFORNIA

In the classic days of Ferrari, many considered it in poor taste to announce the model name or number anywhere on the car. It was a Ferrari, which was enough; if you didn't know what sort of Ferrari, you were not the sort of person who cared.

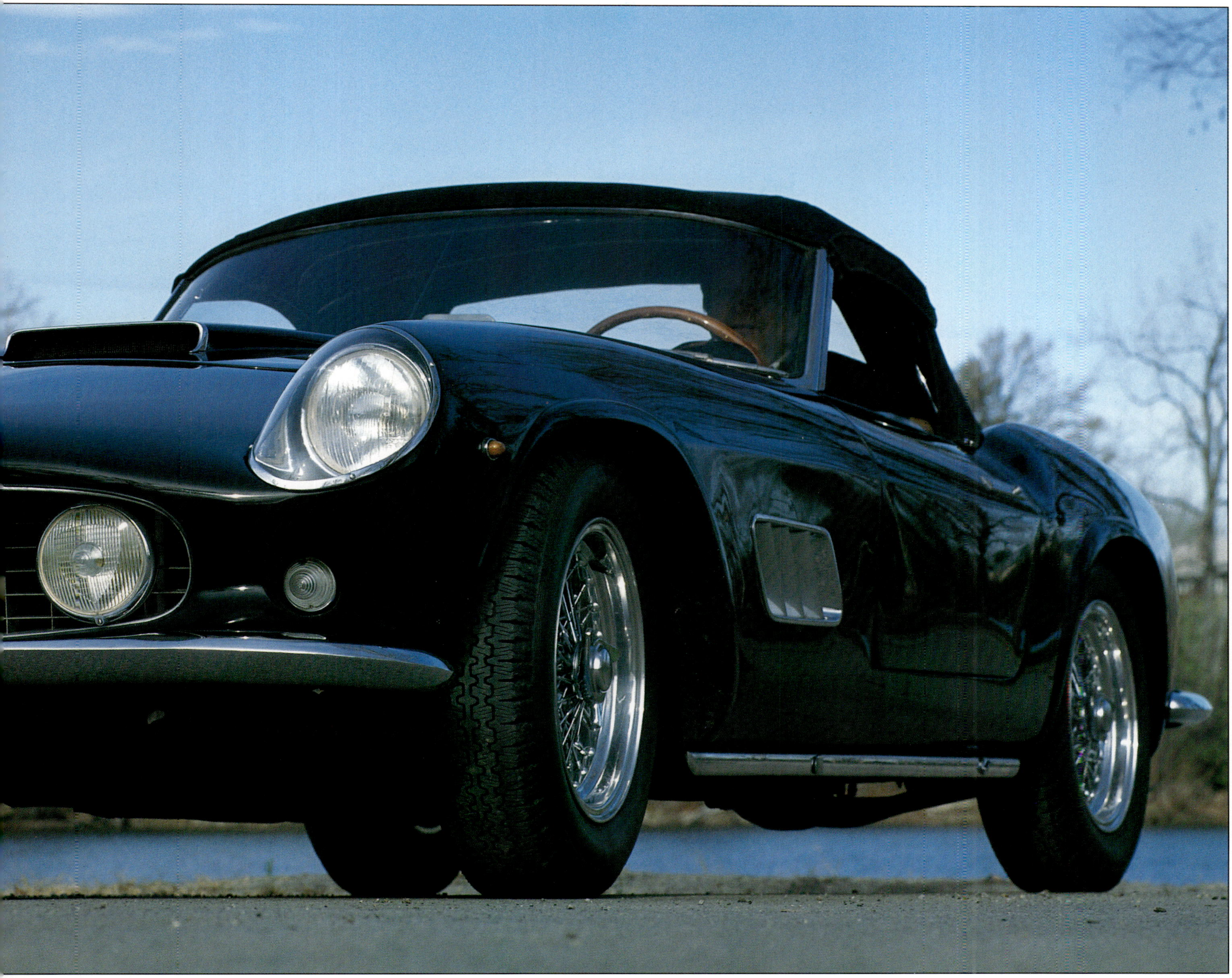

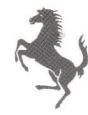

1959 FERRARI 250 GT

In theory, a "cabriolet" – like this one – is a more luxurious and better-appointed car than a "spyder," and strictly an open racer is a "corsa," just as a racing coupe is a "berlinetta." Today, though, the distinctions have been blurred – and in any case, even this well-appointed cabriolet had a very fair turn of speed, as evidenced by the 300 km/h (186 mph) speedometer. It is a mistake – and one of which Enzo Ferrari strongly disapproved – to specify a racing engine for a car which is to be driven only on the road: a slightly downrated engine will be much more tractable and reliable.

47

1959 FERRARI 250 GT

48

49

1962 FERRARI GTO

Great attention was paid to aerodynamics on the GTO (also shown overleaf). The rear "lip" spoiler and the low nose were both designed to increase down-force; the wind tunnel at the University of Pisa was used to refine the design, but the net result was still a car that was very beautiful, where art and science were combined more perfectly than ever before – and arguably, better than ever since. As befits a racer, a huge tachometer dominates the instrumentation.

51

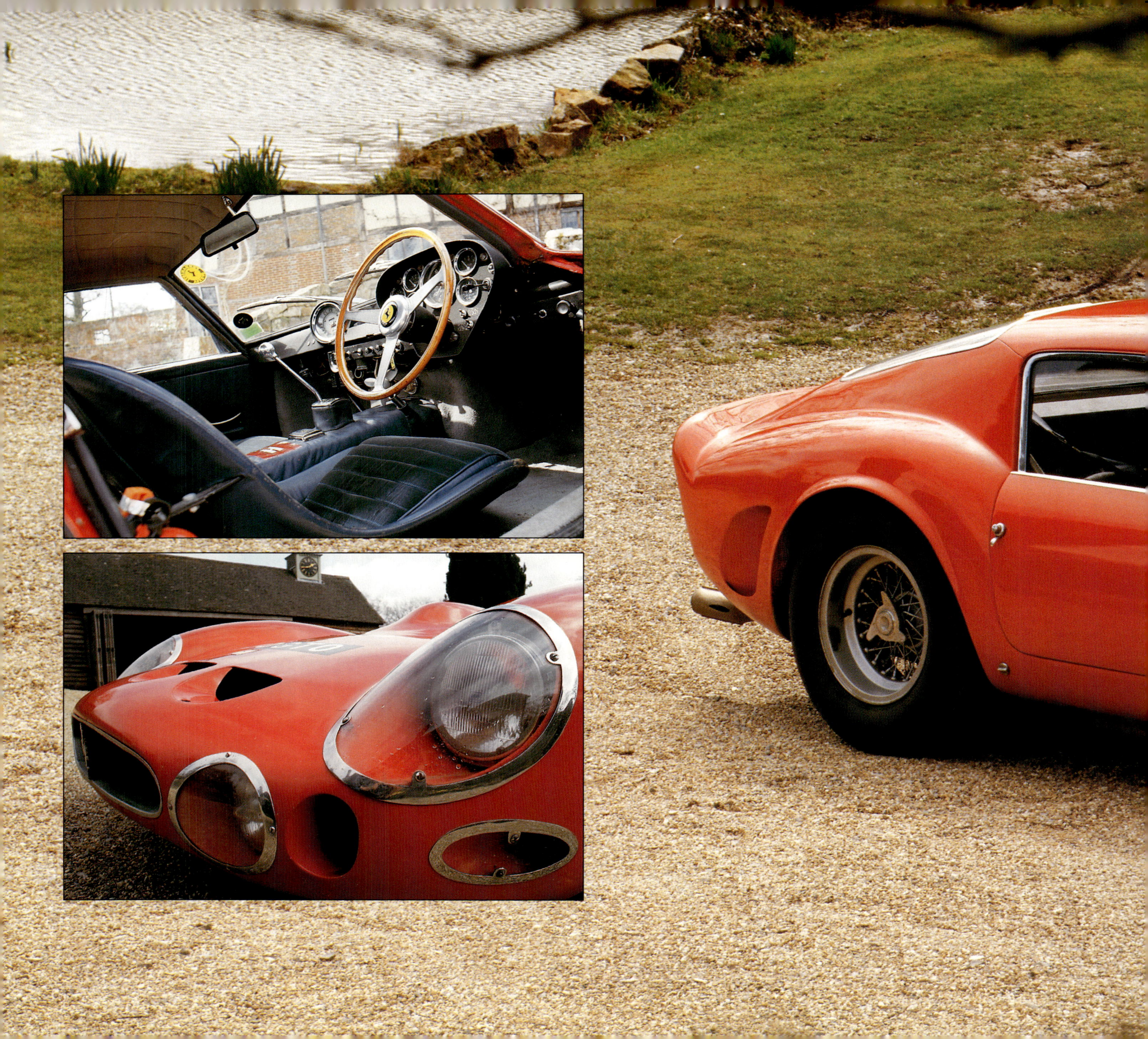

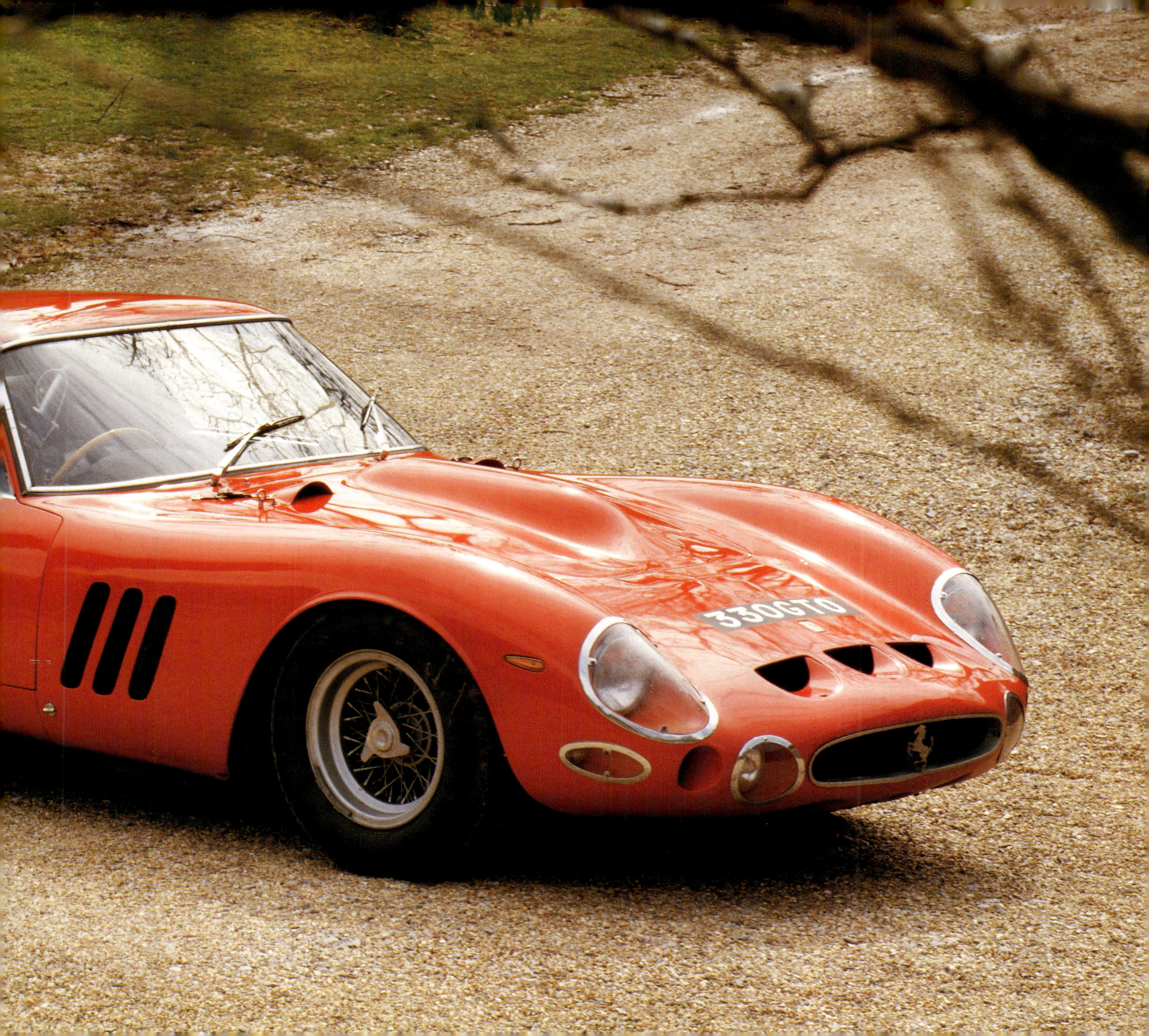
330GTO

1963 FERRARI 250 GT SWB LUSSO

"Lusso" (luxury) and "Berlinetta" (racing coupe) may seem incompatible, but Pininfarina managed to reconcile the opposites very well, in one of his finest designs of all time. The engine was moved forward slightly, compared with the original SWB racers, to allow a more capacious cockpit; the body was of steel, not light alloy; and the fit and finish was appropriate to a luxury car, rather than being built to last just a few racing seasons. GM stylist Chuck Jordan called it "timeless," and who would argue with him?

1963 FERRARI 250 GT SWB LUSSO

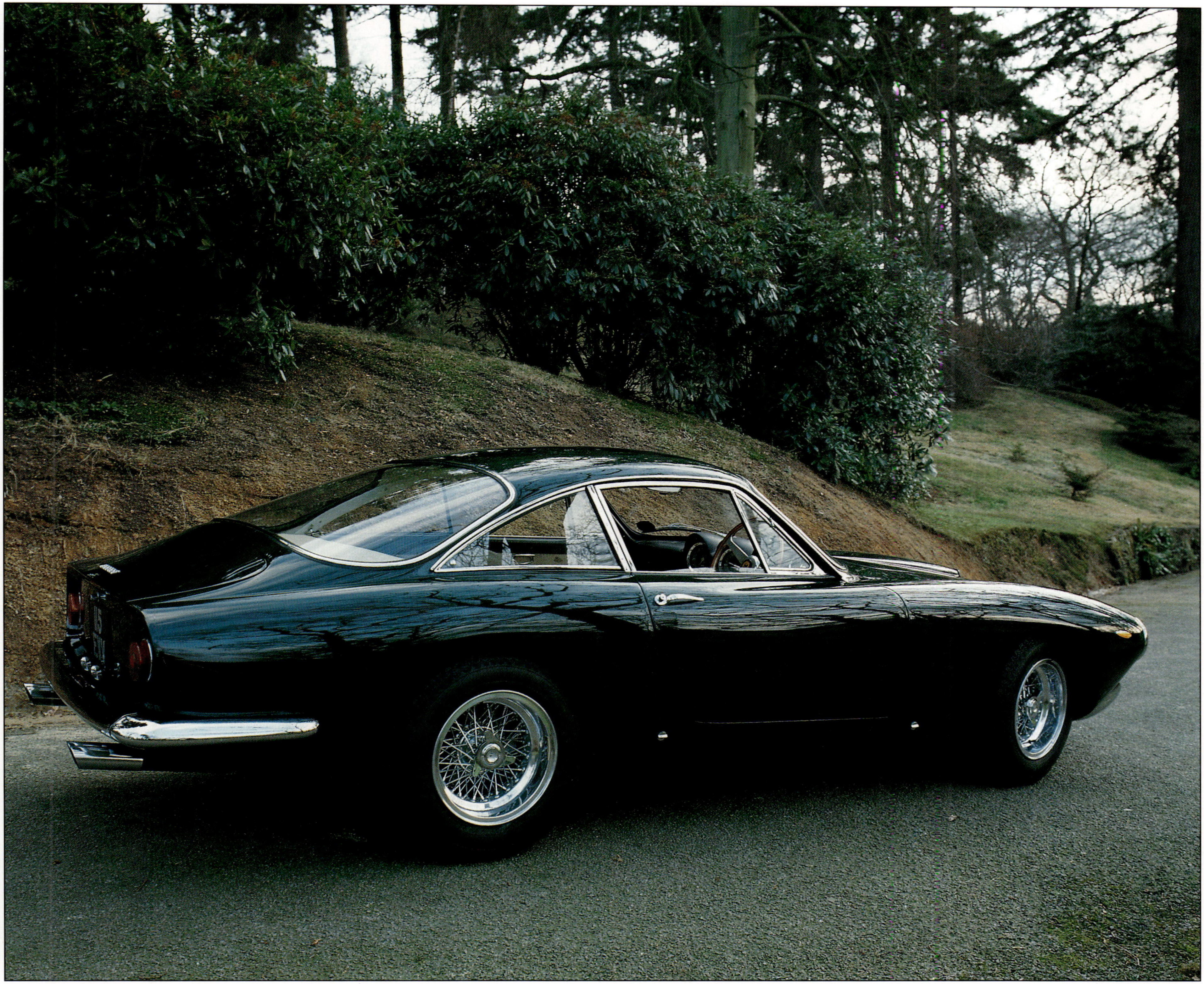

1964 FERRARI 250 GT LUSSO BERLINETTA

The 250GT was discontinued in 1964, and the last cars built (out of a production run of some 2,500) were Lussos very like this one. To date, this was far and away the largest number of (roughly) similar Ferraris that had ever been built, and it is the earliest classic Ferrari that is likely to be encountered on the open market. The Pininfarina body was precisely right, from the very start, and the Borrani wire wheels were very much in keeping with the style of the whole.

59

1965 FERRARI 500 SUPERFAST

"Superfast" was originally a Pininfarina styling exercise, a name applied to a series of four show cars built on just three 400 Superamerica chassis. When the Superfast 500 entered production as a successor to the 400 Superamerica, it was, however, fitted with an even bigger engine than the Superamerica, a five-liter hybrid derived principally from the Colombo design (with detachable heads) but with the bore spacing of the Lampredi block. About 25 Series 1 Superfasts were built from 1964 to 1966, and a dozen Series 2 cars followed. This car (also shown overleaf) is one of seven right-hand-drive models made to be sold in England. All Superfasts were the most luxurious Ferraris obtainable, and they attracted a wide range of customers including the actor Peter Sellers as well as the Aga Khan, Prince Bernhard of the Netherlands, and the Shah of Iran – who bought both a Series 1 and a Series 2.

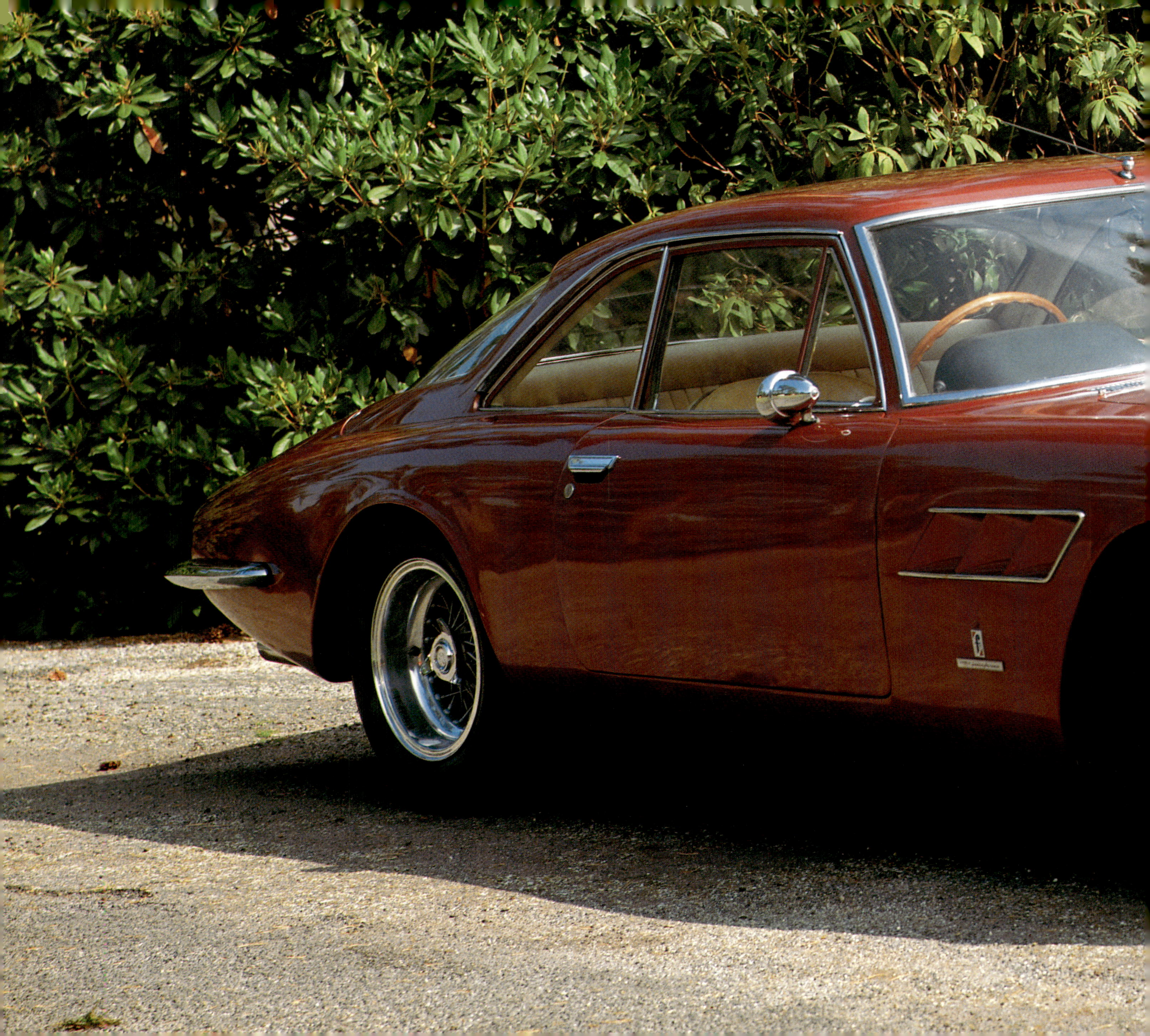

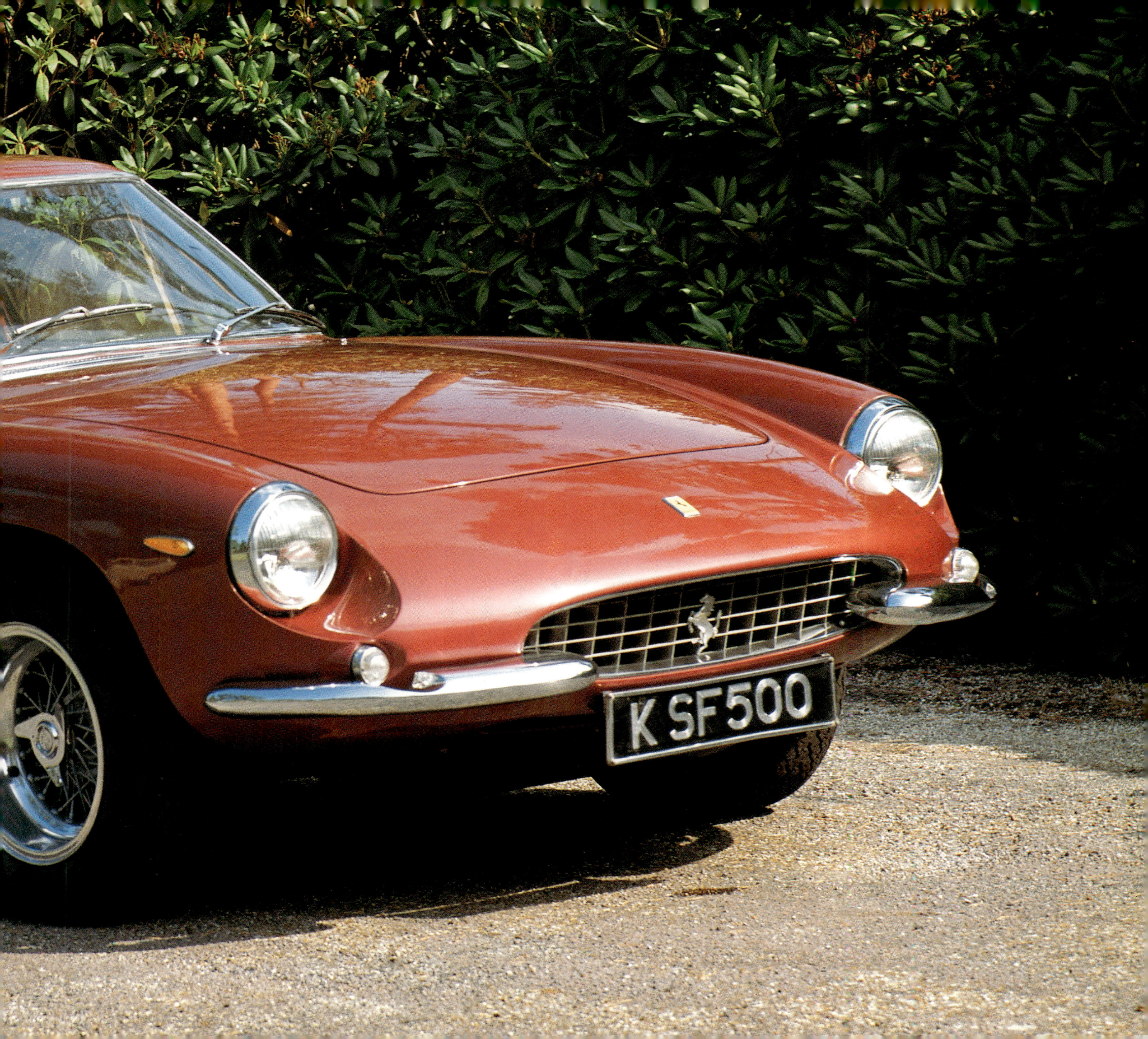
K SF500

1965 FERRARI 500 SUPERFAST

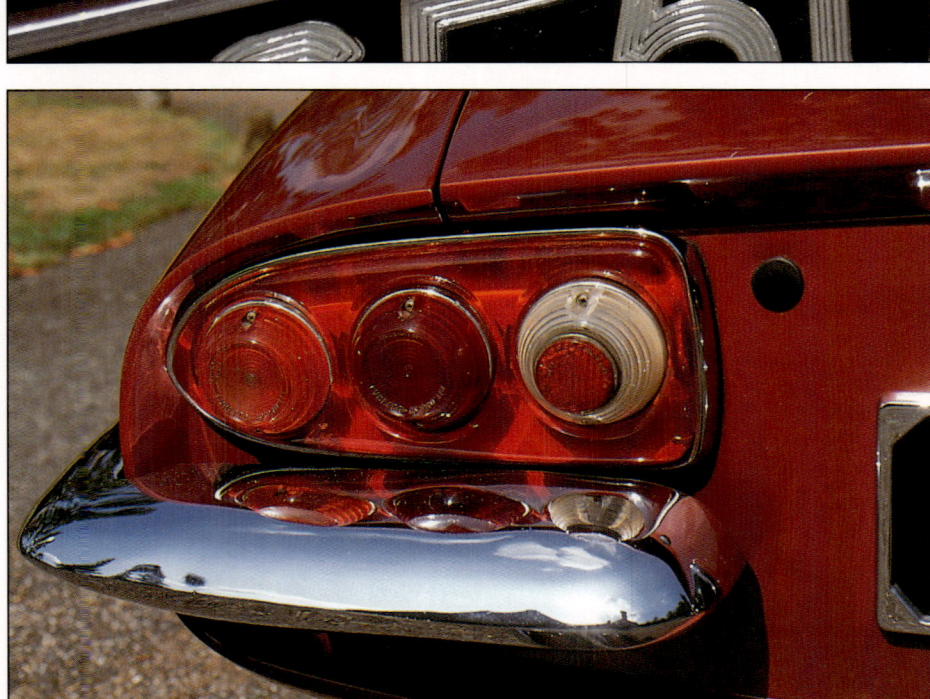

The Superfast is one of those rare cars where you can go on admiring the details almost forever. The only thing that looks particularly old-fashioned today is the combined reversing light/reflector, which has almost a 1950s look to it. The prancing horse is particularly finely sculpted, and even the side extractor vents with chrome on three sides are a thing of beauty. Vents of this type were almost a Pininfarina trademark.

1966 FERRARI DINO 206 SP

The original Dino 206 SP show car was a racer – and it showed. The mid-engined V6 was refined over several shows before actually entering production. It was a very forward-looking car, with alloy wheels instead of wires and a mid-mounted engine, yet paradoxically it represented a return to Ferrari's roots as a builder of small, light, sweet-handling cars based heavily on racing practice. Cars like the 500 Superfast had come a long way from those origins.

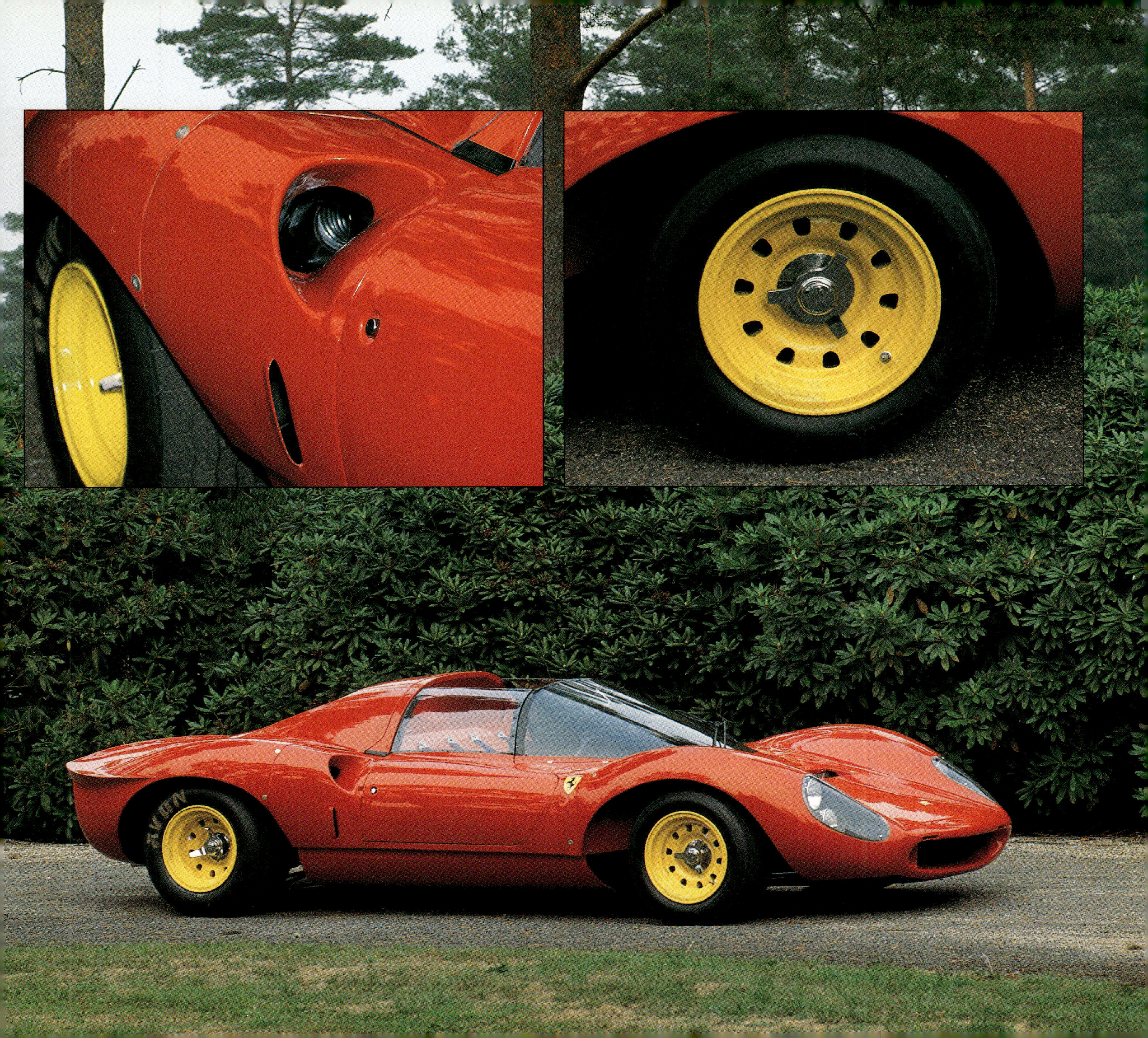
AVON

1966 FERRARI DINO 206 SP

The production version of the Dino would have a somewhat less radical rear window treatment, though still with a strong reverse curve; the tail treatment would be less futuristic, and the ventilation slots in the rear would not be just black mesh. The engine would also be mounted transversely, and much less accessible.

1966 FERRARI DINO 206 SP

The absence of a transmission hump meant that the seats could be very close together (though the Dino was not really a car for canoodling in), but the enormously wide sills on the SP prototype made getting in and out an interesting exercise. The car was built on a racing chassis, and delightful though the 206 was when it came out, there were those who wished that it had been even closer to the prototype! For the production car, the single enormous racing-style pantograph wiper was replaced with twin wipers, but in a curiously retrograde step the twin lights were replaced with single units each side. It is not clear why Enzo chose not to call the final production car a Ferrari. Was he trying to establish a new marque, in honor of his dead son, or was he simply unhappy about putting "Ferrari" on a car which had an engine designed by Ferrari engineers but built by Fiat? The same engine was used in the front-engined Fiat Dino.

1967 FERRARI 275 GTB/4

The 275 GTB/4 (also overleaf) was the first roadgoing Ferrari to be fitted with the four-cam version of the Colombo V12, delivering 300 bhp at 8000 rpm. The interior was spacious, well-appointed, and as easy of access as such a low car could be: an advantage of traditional chassis construction over monocoques.

73

1967 FERRARI 275 GTB/4

74

75

1967 FERRARI 365 CALIFORNIA

The 365 California GT seen here and overleaf is extremely rare (only 14 were built in 1966 and 1967), and it was unlike previous Californias in that it was, with its 4.4-liter engine, more like a Superamerica or Superfast. It was definitely the most luxurious California, and it is regarded by many as the last of the old-style big-engined luxury Ferraris.

76

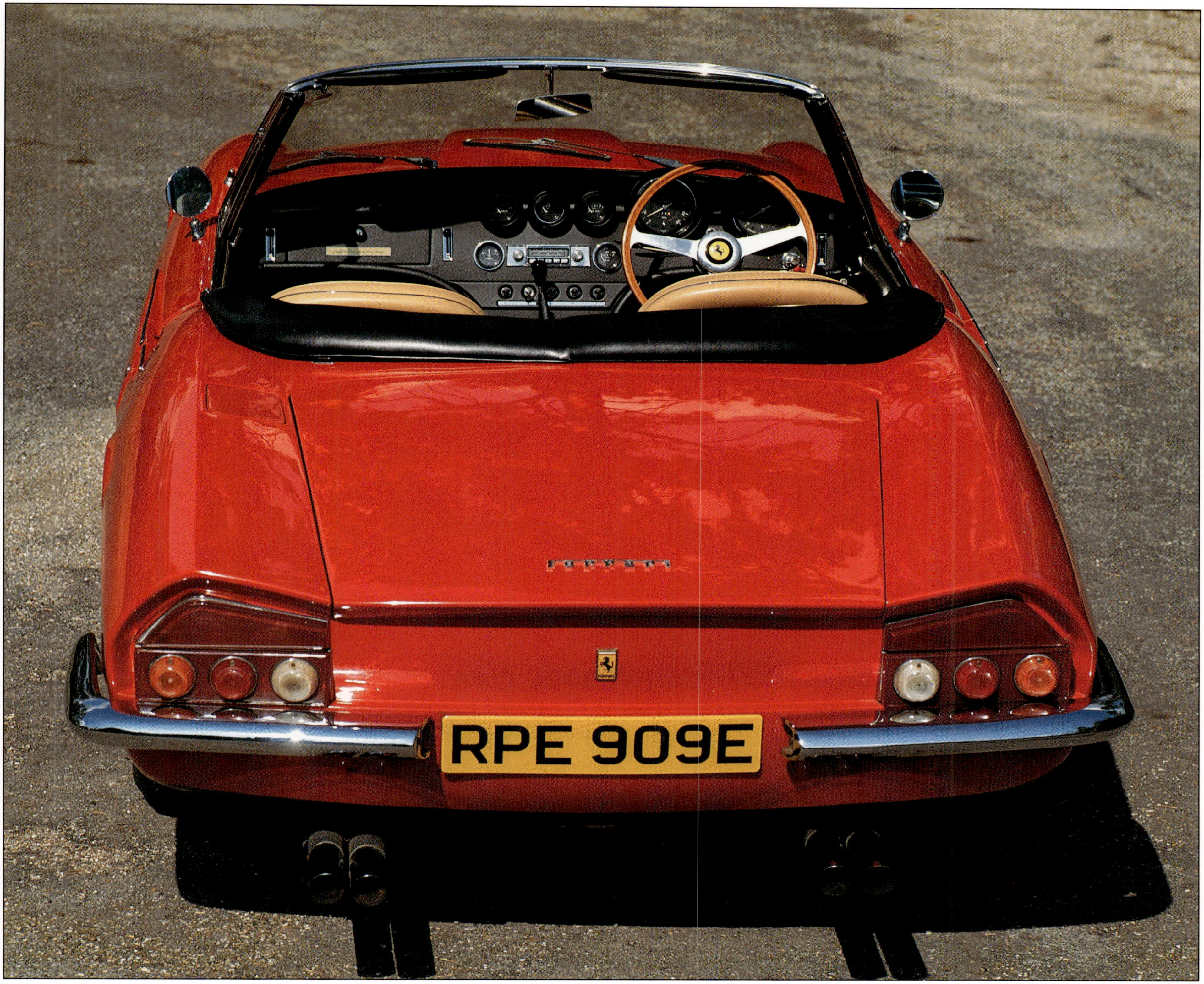
RPE 909E

1967 FERRARI 365 CALIFORNIA

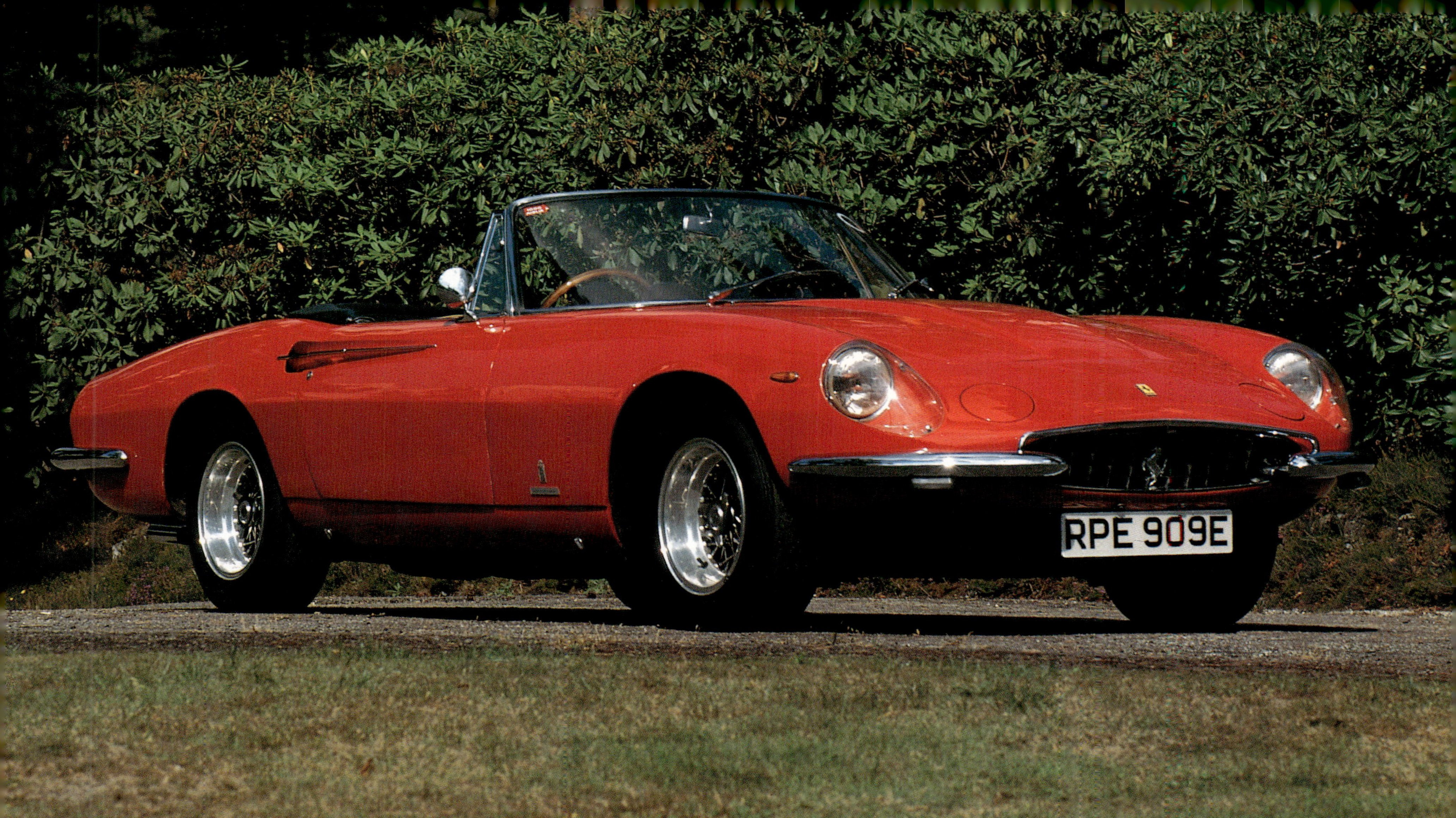
RPE 909E

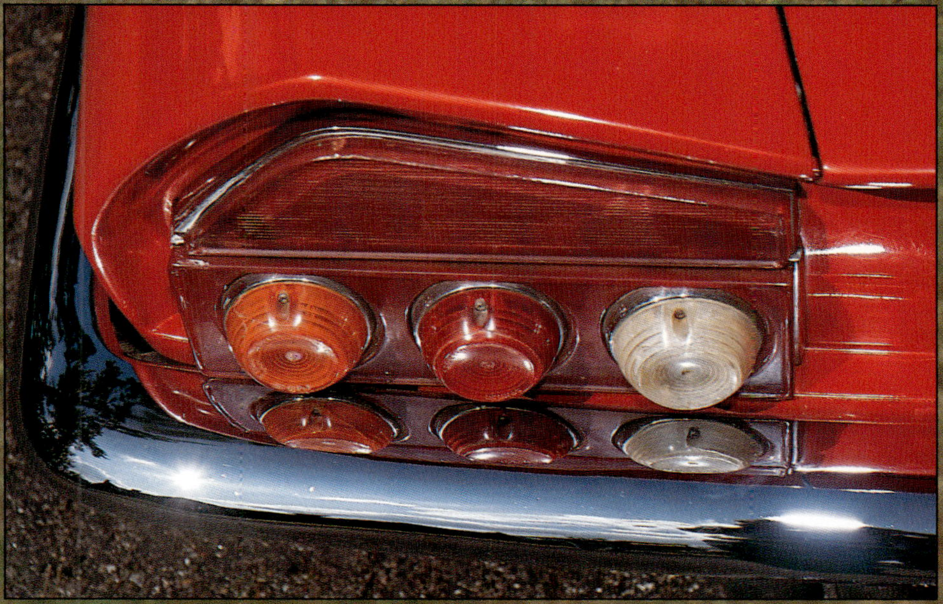

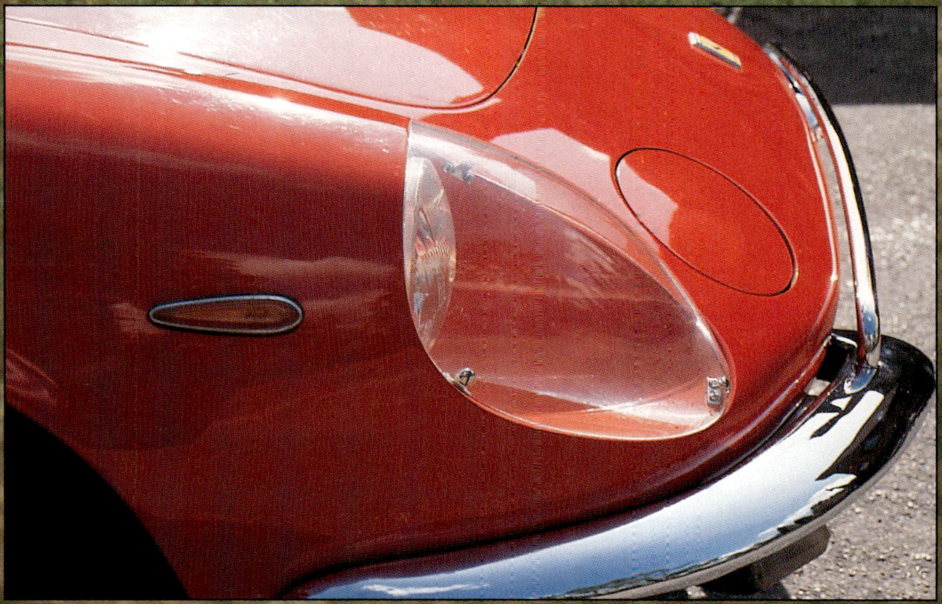

1968 FERRARI 275 GTB/4

Production of the GTB/4 was effectively halted by Federal emission control regulations in 1968, after only about 280 cars had been built: it would have been impossible to bring the 3.3-liter engine into compliance with the laws that came in that year, without reducing the power and efficiency so far that no-one would have bought the car. The GTB/4 was, in a sense, the last of the "old" Ferraris. Sure, it had acquired disc brakes and a four-cam engine and all kinds of other things that were very different from the cars of the 1950s, and it was an evolution of the enormously successful 250GT series, but in future Ferrari was going to have to pay more and more attention to legislators throughout the world who thought that they knew more about building cars than Enzo Ferrari did. Also, the lithe yet rounded body of the 275 GTB/4 was recognizably derived from its predecessors, the three-liter cars; there would never be another Ferrari in quite the same style.

1972 FERRARI 365 GTB/4 DAYTONA

By the 1960s, a racing car was so specialized that it could not really be used on the road: the day of the true sports/racer was passing fast. What Ferrari did, therefore, was to apply racing lessons to a car that was frankly too heavy and luxurious to be raced: a very fast roadster, rather than a racer. It was raced, and successfully, even at Le Mans; but the sheer weight of a 3600-lb. car meant that brake fade was always a problem.

1972 FERRARI 365 GTB/4 DAYTONA

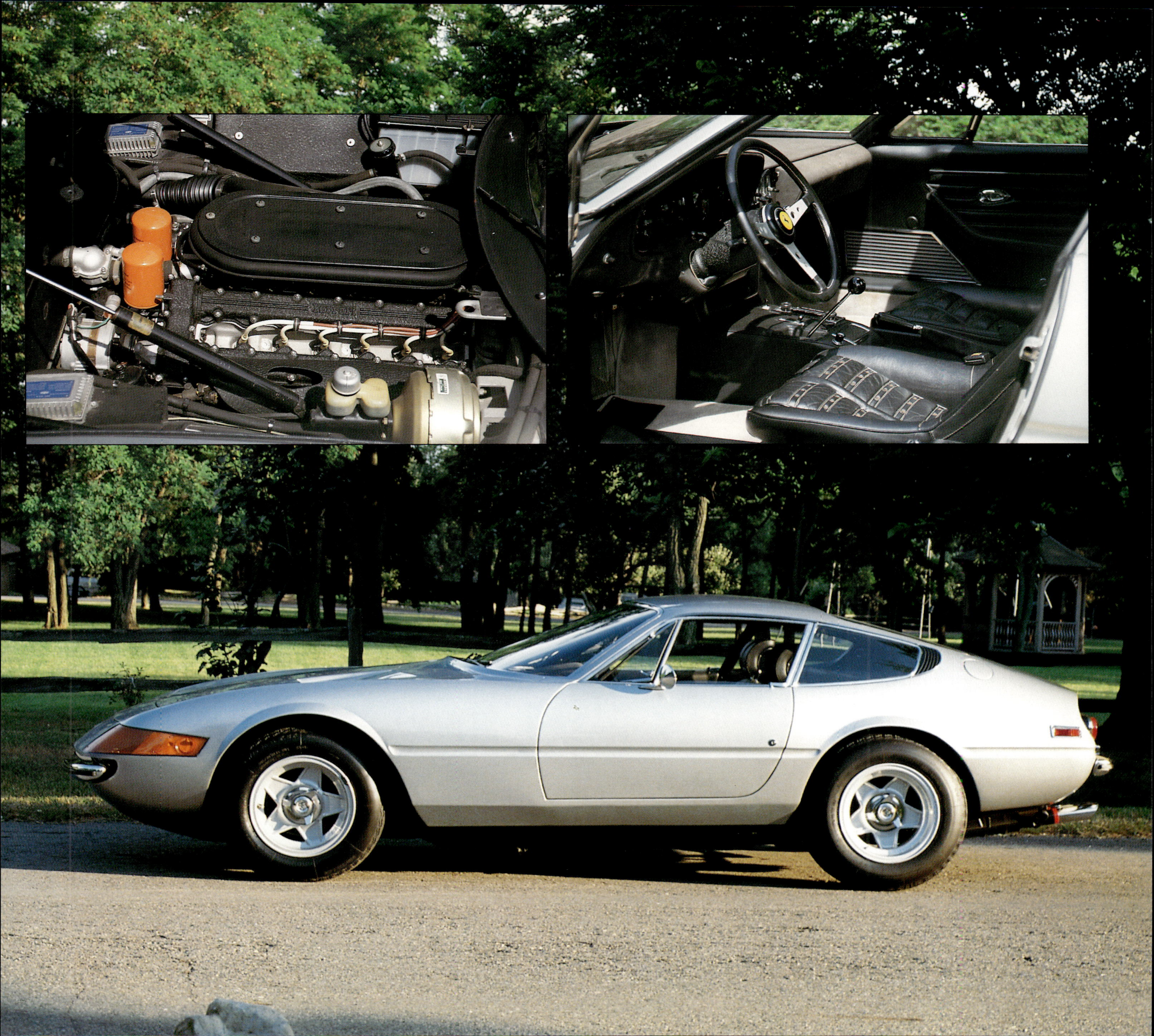

1973 FERRARI GTS/4 DAYTONA SPYDER

The vast majority of the 1300 or so Daytonas were berlinettas (racing coupes), but a good number of "Daytona Spyders" were also built – about ten per cent of the total. Several berlinettas have since been converted into cabriolets.

86

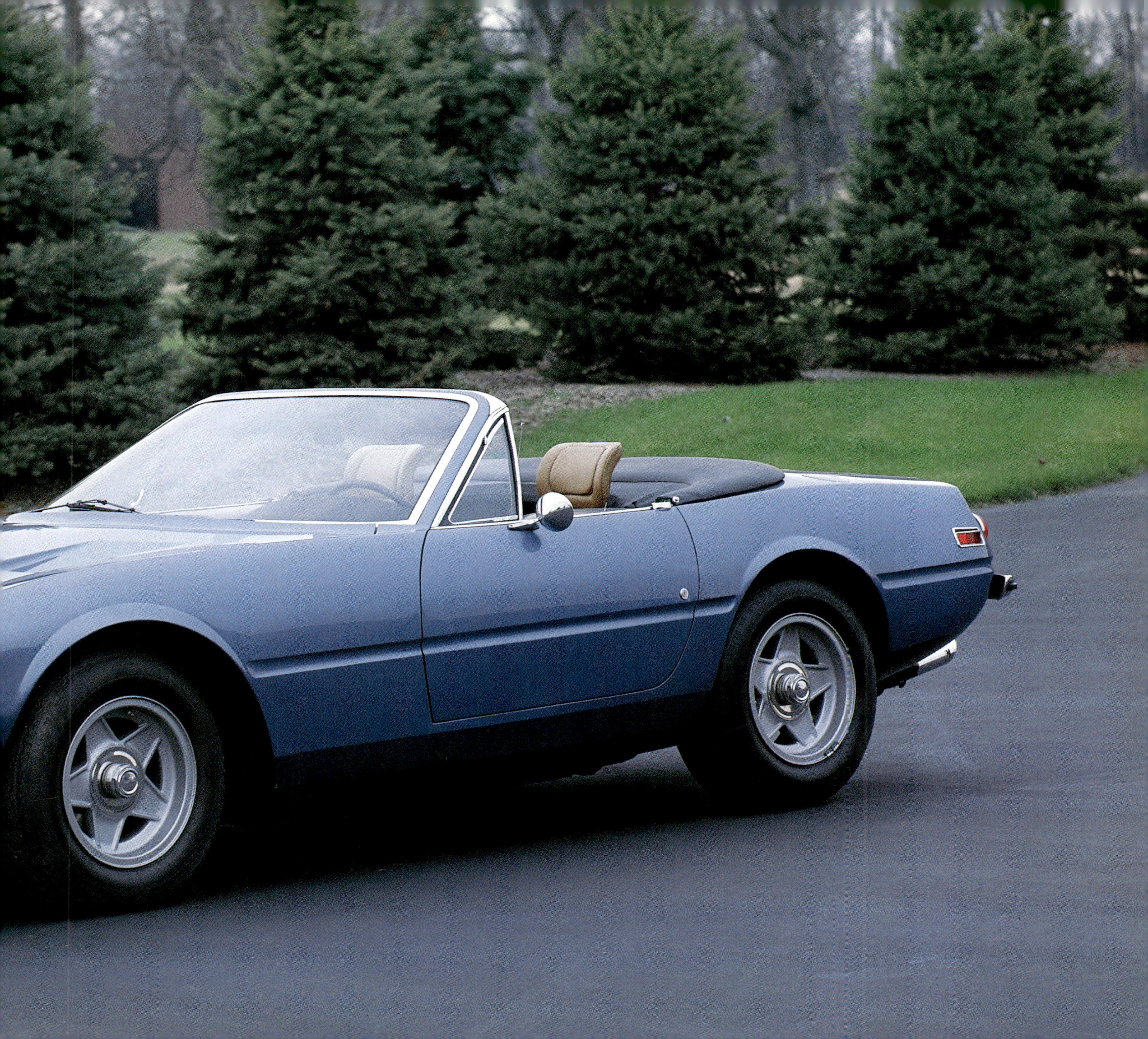

1973 FERRARI DINO 246 GTS

88

About 4,000 iron-block 246GT Dinos were built, making them very much more common than the hundred or so alloy-block 206GT cars. The Targa-top GT Spyder shown here and on the next two pages accounted for about 1,200 cars.

1973 FERRARI DINO 246 GTS

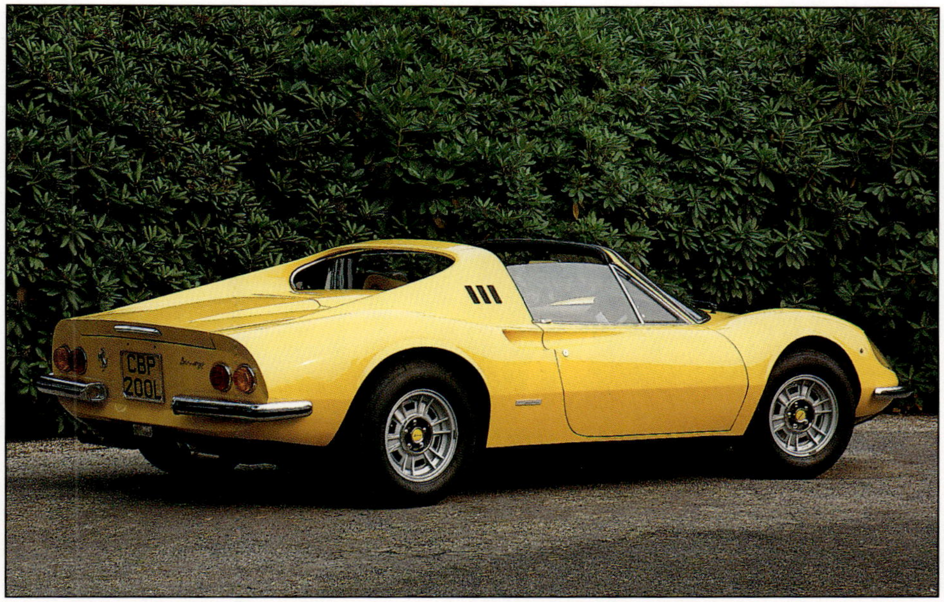
CBP
200L

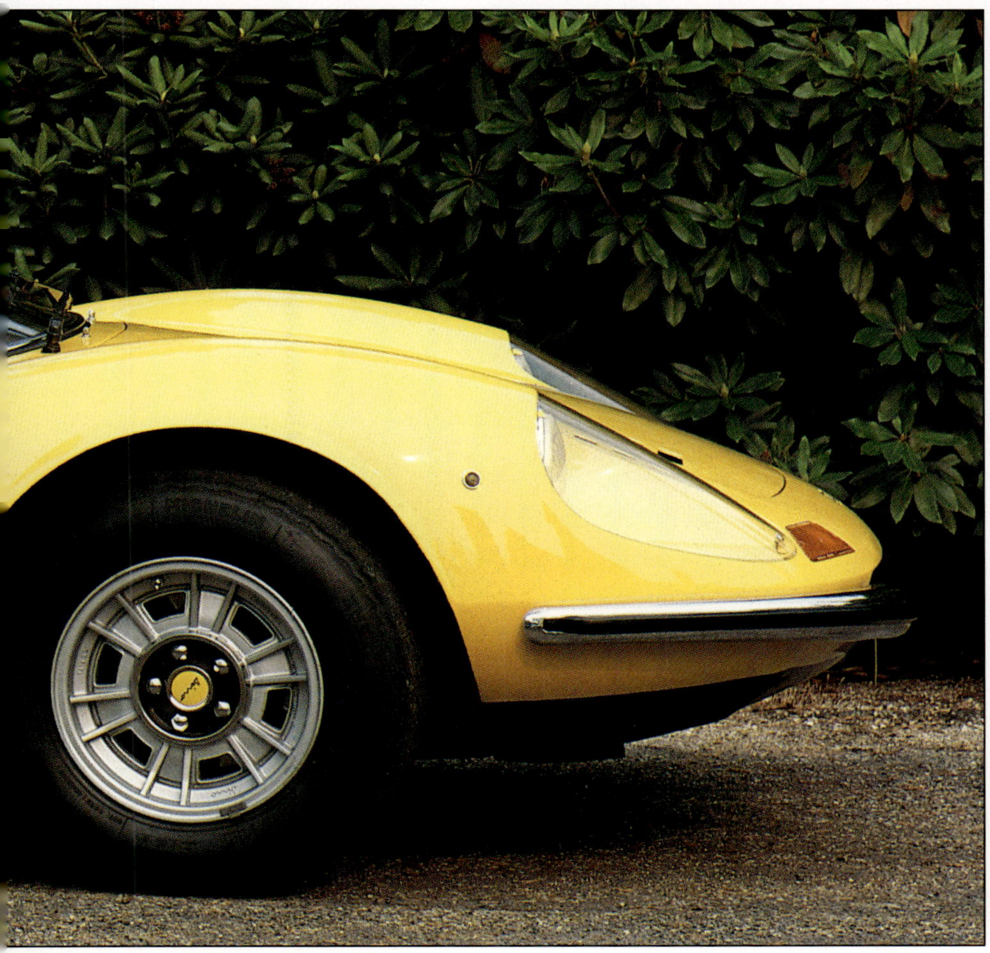

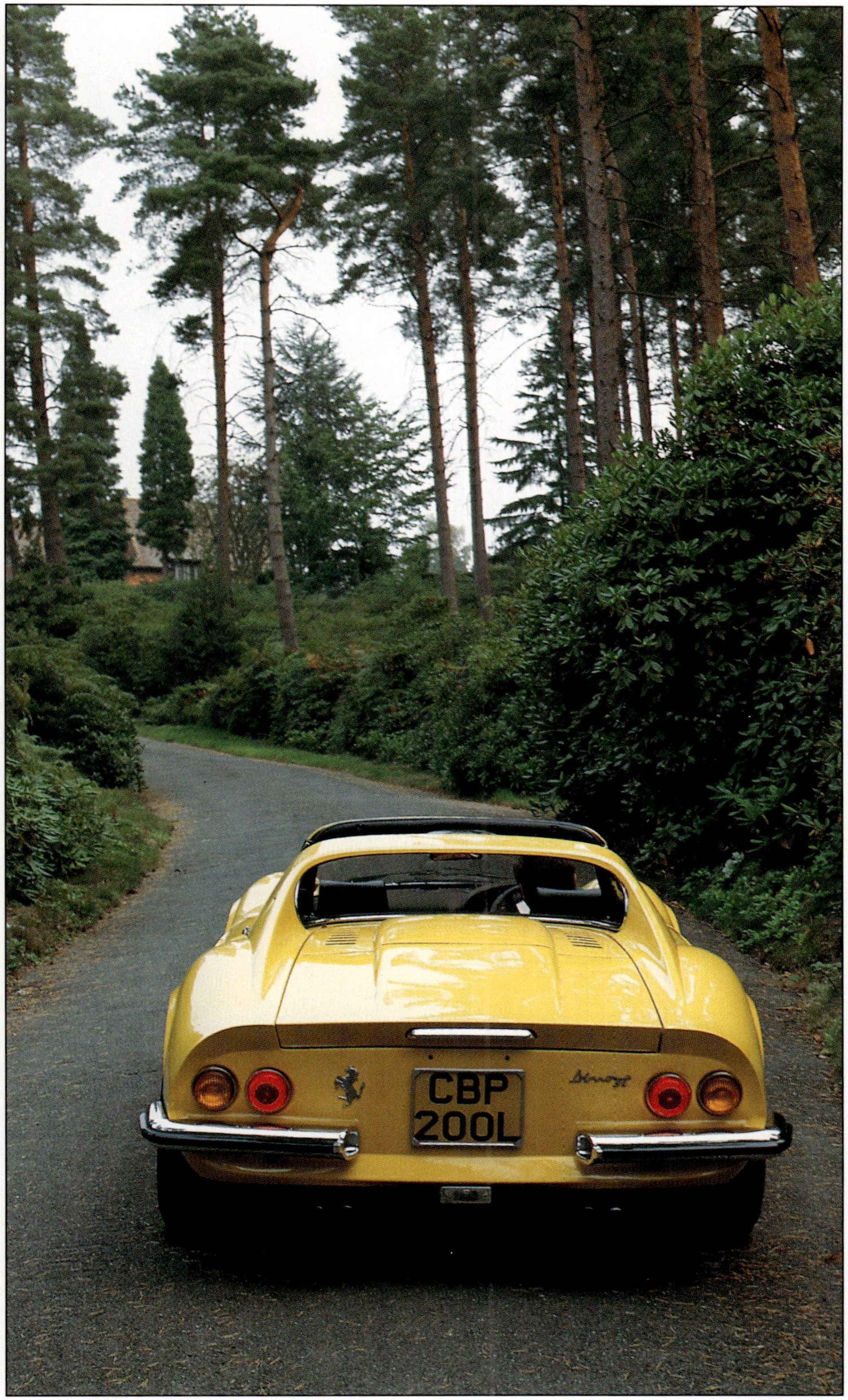
CBP
200L

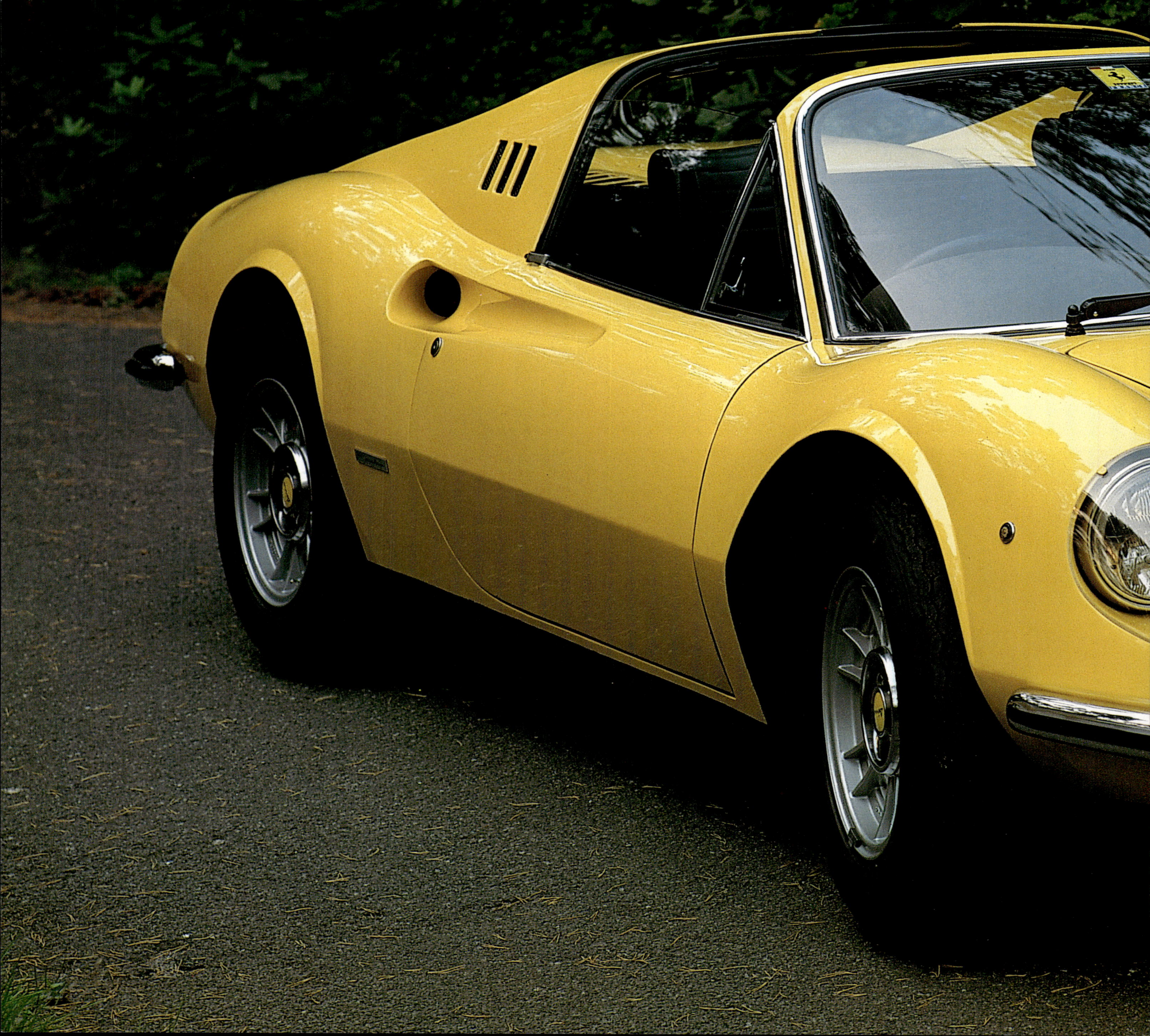

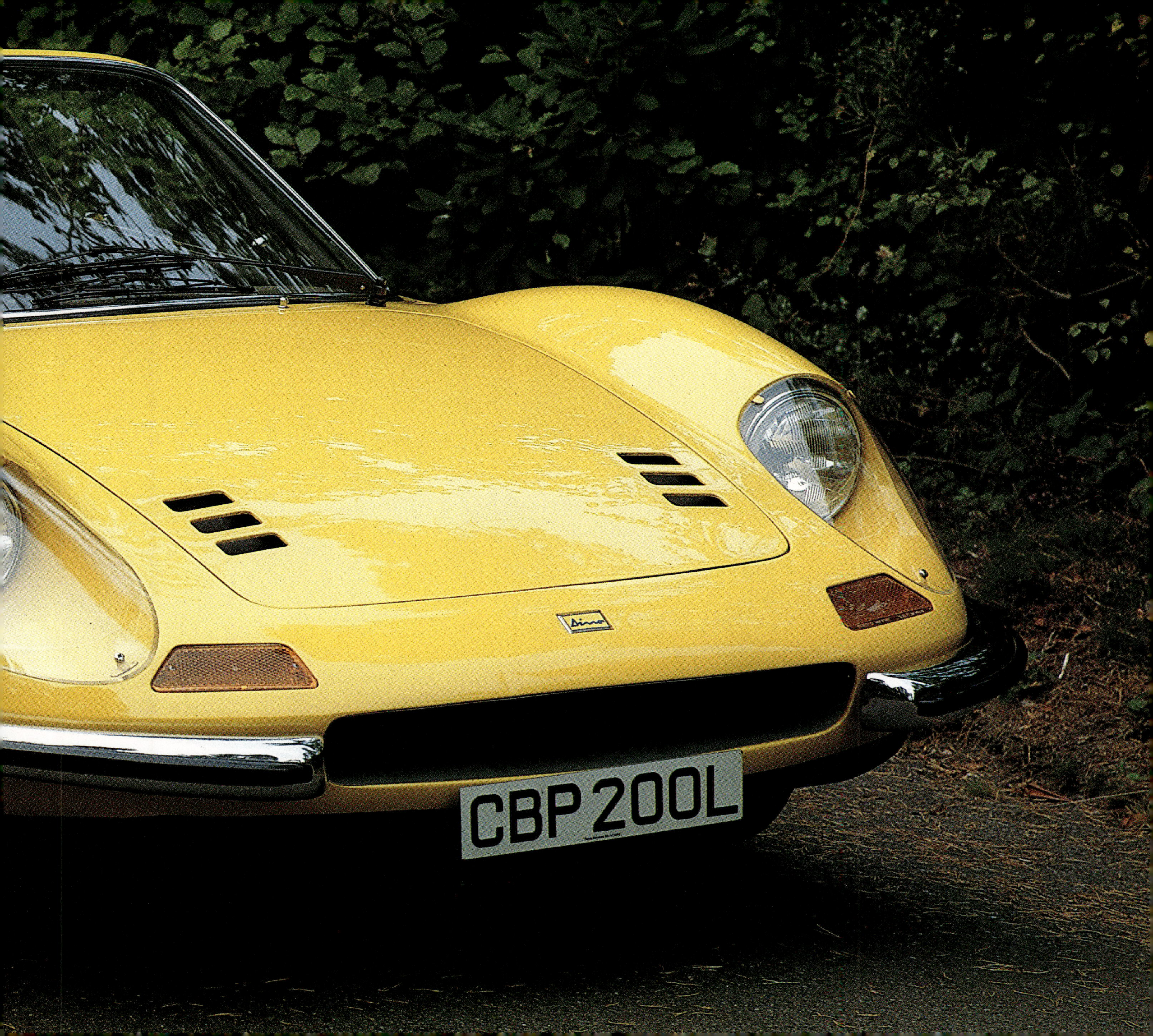
CBP 200L

1979 FERRARI 512 BERLINETTA BOXER

The strikingly handsome Berlinetta Boxer bore a family resemblance to the Dino but was clearly a much bigger, more powerful car, with the full complement of cylinders expected of a Ferrari: twelve.

1979 FERRARI 512 BERLINETTA BOXER

The five-liter engine of the original Berlinetta Boxer was derived from the Grand Prix engine, and the 512BB was a larger version of the same thing. It is a very neat unit, but the extremely low lines of the car mean that accessibility remains a problem. Of course, given the kind of luxury that it had to propel, it needed to be a big engine! The engine-deck treatment, with the twin covers, would be greatly improved on the next flat-12, the Testarossa.

1984 FERRARI QUATTROVALVOLE 308

The 308 quattrovalvole (four valves per cylinder) was another development of Ferrari's mid-engined theme. They seemed to have lost their way, though, with V6, V8 and flat-12 configurations, but no mid-engined V12 cars.

1986 FERRARI TESTAROSSA

The Testarossa was outrageous – but magnificent. Huge and immensely powerful, it boasted a 200 mph speedo and justified it with a top speed of 180-plus mph.

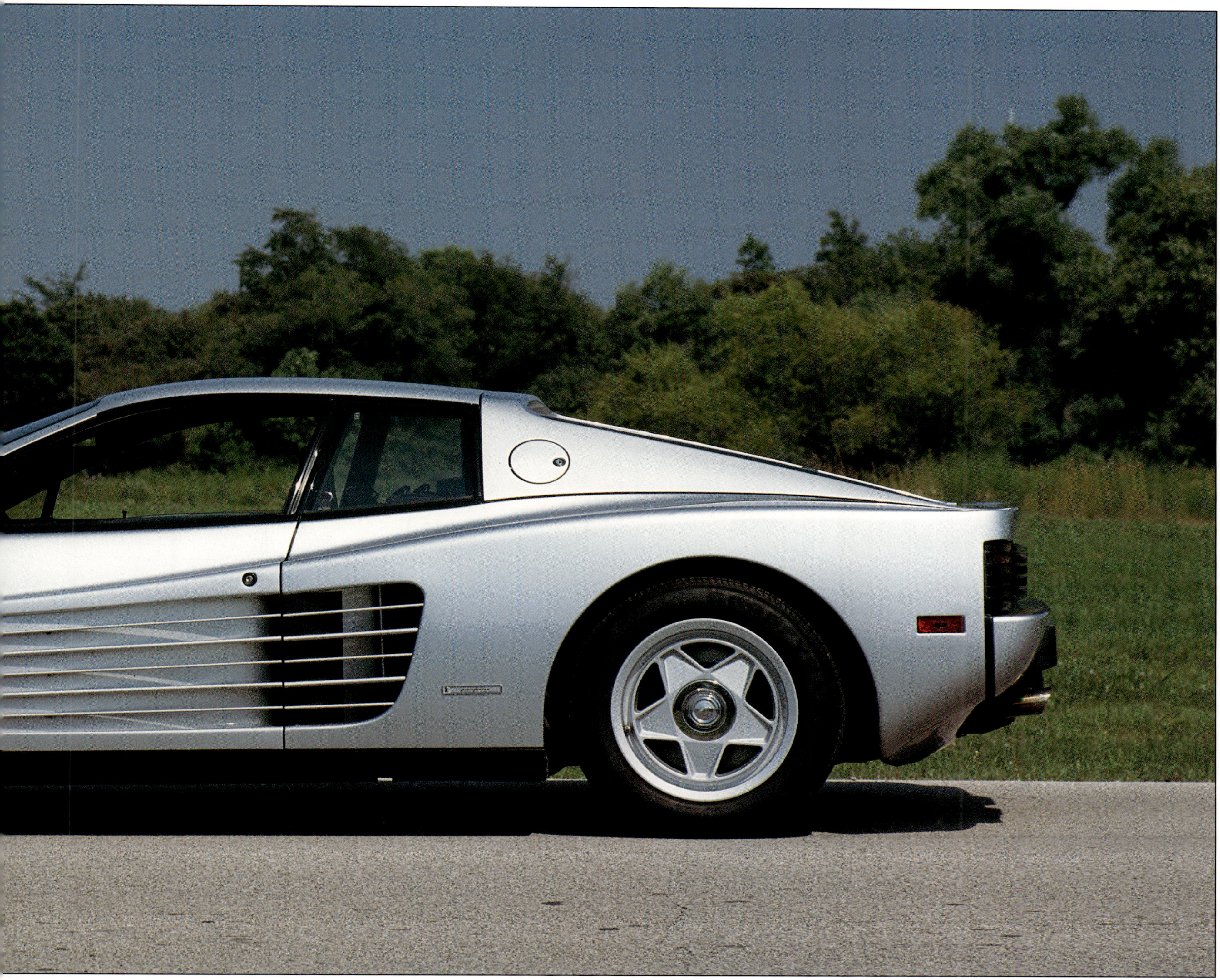

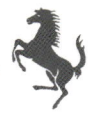

1986 FERRARI TESTAROSSA

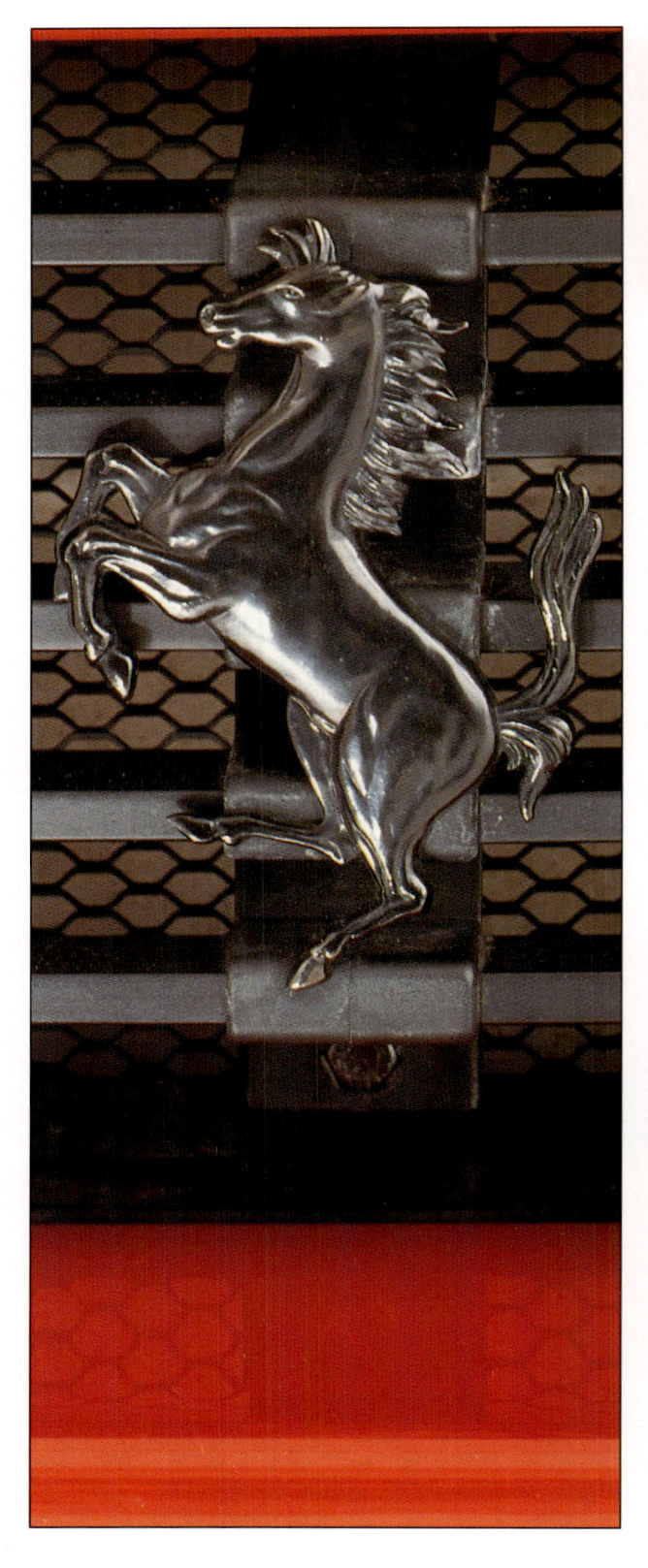

The widest roadgoing Ferrari built, the Testarossa was six inches wider than a BB and was exquisitely detailed. The engine deck treatment was a masterpiece of design, and underneath it lurked a 380 bhp version of the by now well established flat-12. It was definitely a return to the days when the mere sight of a Ferrari stopped people in their tracks.

testarossa

1989 FERRARI F-40

104

To celebrate forty years of the marque, Ferrari built the F40: a stunning combination of good looks and sheer, unadulterated power. The massive composite wheels, thirteen inches wide at the back, were retained by hub-spline nuts and retaining clips; the body was made of ultra-light Kevlar and carbon-fiber reinforced plastic; and on the earlier models, there were absolutely no concessions to even basic luxuries such as the wind-up windows seen here. Originally, they were just sliding plastic. The distinctive triple exhaust (no mean trick from a V8!) comes from a massive collection/expansion chamber above and behind the engine, but most people would not even notice it – they would be mesmerized by the "wing" instead.

1989 FERRARI F-40

Which would you rather have: the car or the house? Lifting the entire rear of the car reveals the massive engine, and shows that the actual rear window separates the passengers from the engine compartment.

107

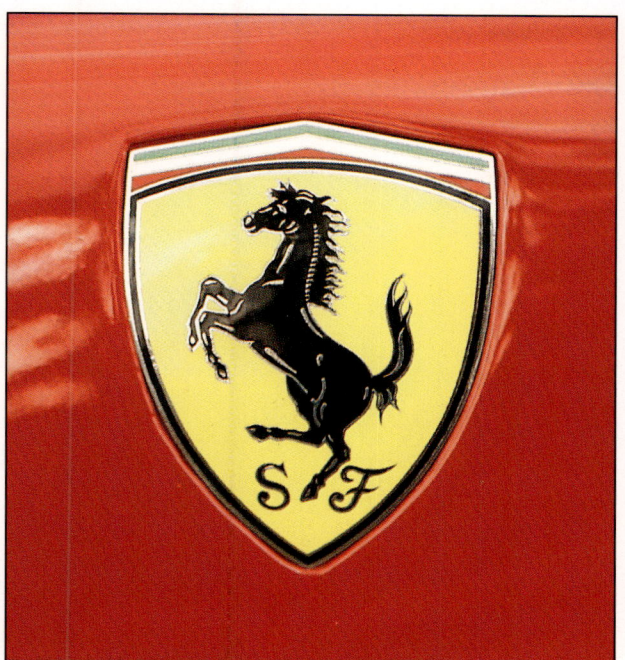

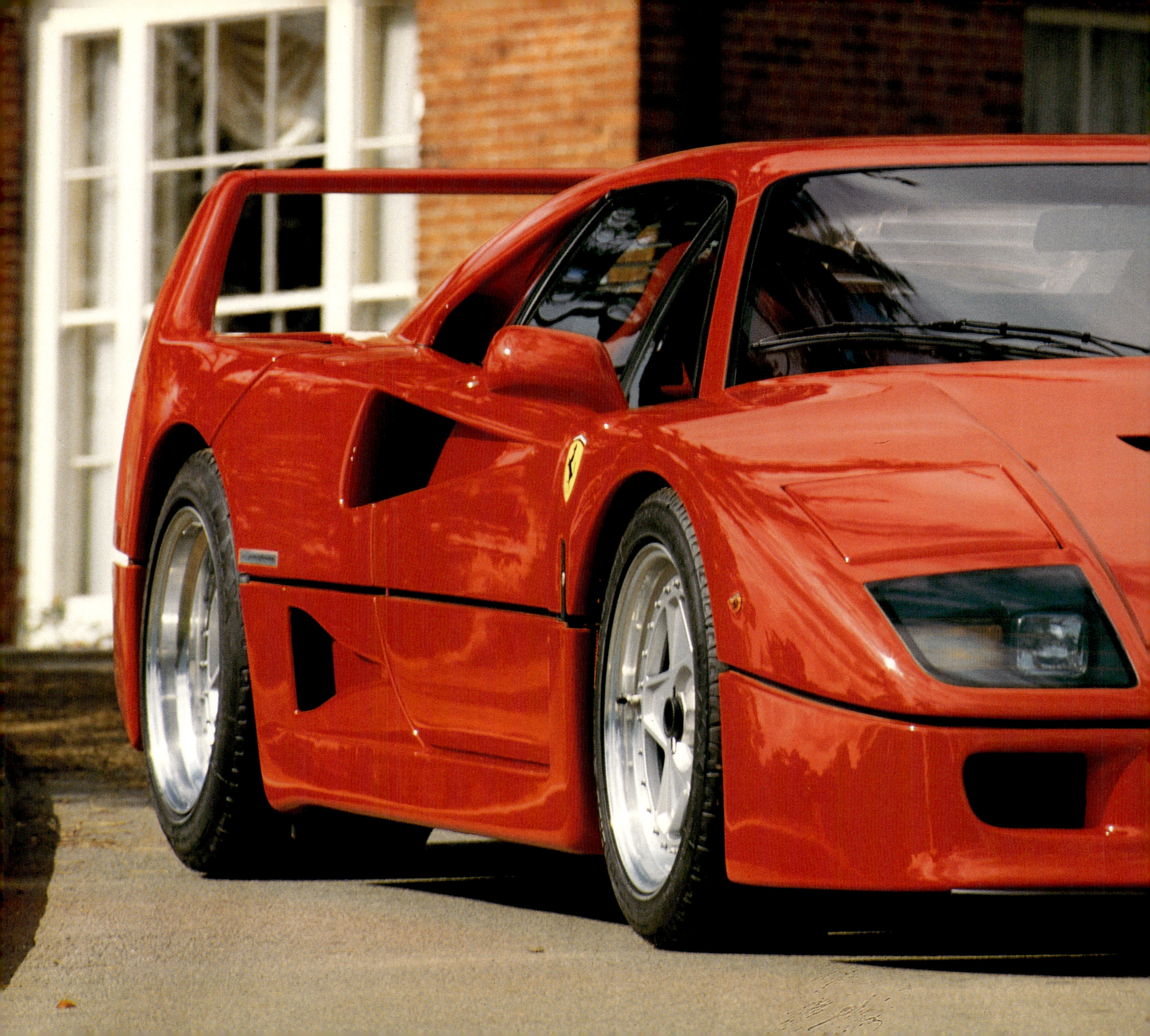

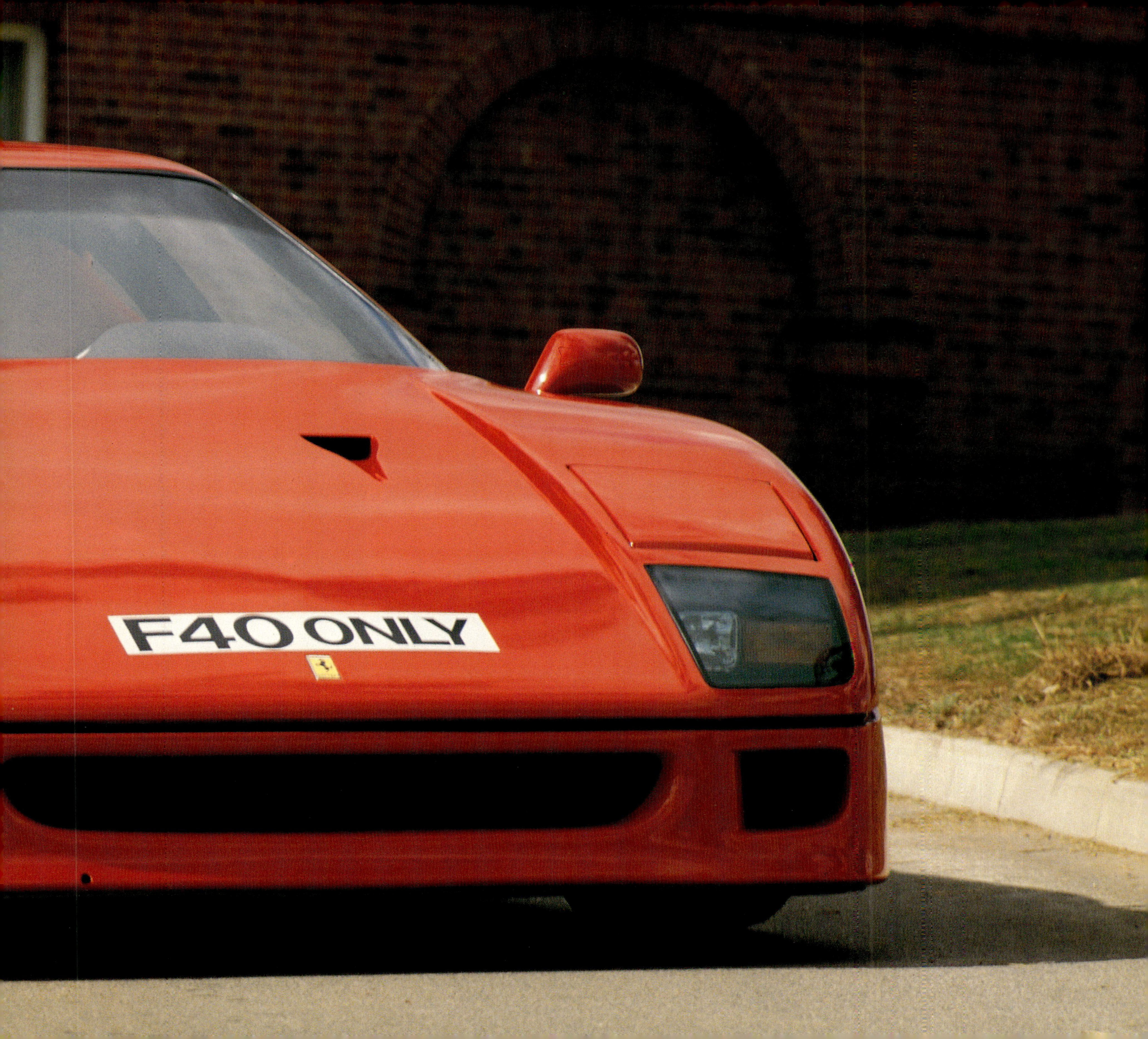
F40 ONLY

1992 FERRARI 512 TR

With the F40 out of production, the 512TR (TR for Testarossa) is Ferrari's flagship. Slightly more restrained than the original Testarossa, it shows how Ferrari has moved in many ways into the mainstream – look at those five-stud wheels, for example – while still retaining the power to turn heads and to leave no-one in any doubt about who built the object of their admiration. Just as they would have done twenty or thirty or forty years ago, enthusiasts still say (rather quietly, and reverently) "It's a Ferrari!"

1992 FERRARI 512 TR